PRAISE FOR *HERE TO HELP*

"Kevin Thorn is a young entrepreneur ~~who has~~ embraced the new revolution in pest c~~ontrol... who knew~~ there is a revolution—but it is out there! It's a new way of approaching pest problems with a scientific and environmentally responsible vision. We all want to live in a cleaner world and leave our children a healthy planet. But in order to do that, we must increase crop yields to feed greater populations, control insects that carry disease, and protect the food supply at every step before it reaches our dinner tables.

In Here to Help, Kevin explains the who, what, when, where, and why of pest control to help you know what to look for. His book provides you with a road map to put you ahead of what we know is coming. This can help you get in front of your competition, save you money, and keep you out of trouble. Kevin really is Here to Help!"

–Lloyd Smigel and Pat VanHooser, *pest management consultants and owners of Discovery Retreats*

"As someone who is engrossed in guest service, you look for the best in everything. Why not apply that to every aspect of guest service and make sure your guests, staff, and property are well taken care of and pest free? Finding others who share the passion for guest service is rare enough. Finding someone who applies this same level of passion to services in pest management seemed like a long shot, but when Kevin and I first met, I knew we had someone who would do anything to provide the best in education, application, and delivery of this needed in our industry. His joy and passion is self-evident, and the length to which he cares for each residence, employee, and customer is unparalleled."

–Jessica Rodriguez,
Deer Valley Lodging

"I have had the opportunity to work with Kevin Thorn for several years. His knowledge and understanding of pest control is outstanding. He has a great knowledge of bugs, insects, and rodents and how to control them and is able to do that in a way that is understandable to his customers. Through the worst of infestations, Kevin not only took care of the problem, but he also calmed the fears and concerns of the customers and gave them peace of mind."

–**Ross Whitaker,**
City of Salt Lake

"Kevin Thorn is a bug genius! Our company manages over four thousand units in Utah and had previously searched for a bug guy that was half as knowledgeable and that could really solve our pest problems. Kevin did that for us without his personal interests in mind, with only the goal of providing solutions to our problems. We wouldn't recommend anyone else!"

–**Wendy Pickering**, *president,*
EMG Management

"I first met Kevin Thorn when he responded to a bed-bug problem in one of the college's residence hall suites. I was a bit stressed because the college had never experienced a bed-bug problem. His calming and reassuring approach helped me process better the bed-bug elimination plan that he was explaining to us. It was clear Kevin knew what he was doing!
Kevin Thorn is the consummate professional. He takes his pest treatment career seriously and is eager to educate his clients (and the general public) about their current or potential pest problems.

I would also like to give adulations to Kevin's trained bed-bug detecting beagle, Radar, who stole the hearts of many people here at Westminster . . . It was amazing to watch this animal work. You could tell Kevin really cared about Radar's well-being and held the utmost respect for his 'partner.'"

—**Richard Brockmyer**, *managing director of plant and facilities operations, Westminster College*

"What an immensely helpful book for those that are finding themselves continuously challenged in seeking the best pest solution! I certainly felt like one of those people years ago and have been extremely fortunate and relieved to have been able to work with Kevin. After the expensive and painful process of vetting several pest control companies, I can say with complete confidence that Kevin's expertise, praiseworthy business ethics, and high-quality service surpasses them all. I am a total fan and loyal customer for life!"

—**Carina Lyons**, *vice president, Concept Property Management, Inc.*

"I have worked with Kevin Thorn for many years and have always found him to be professional in all aspects of his work. Kevin is always informative and honest in his analysis and his recommendations. I would recommend Kevin to anyone seeking advice on pest control."

—**Frank Sitton**, *associate director for facilities, University of Utah*

HERE TO
HELP

HERE TO
HELP

PEST MANAGEMENT SOLUTIONS FOR
COMMERCIAL PROPERTIES

KEVIN THORN, ACE

Published by Advantage, Charleston, South Carolina.
Member of Advantage Media Group.

ADVANTAGE is a registered trademark, and the Advantage colophon is a trademark of Advantage Media Group, Inc.

Printed in the United States of America.

ISBN: 978-1-59932-728-0
LCCN: 2016938543

Book design by Katie Biondo.

This publication is designed to provide accurate and authoritative information in regard to the subject matter covered. It is sold with the understanding that the publisher is not engaged in rendering legal, accounting, or other professional services. If legal advice or other expert assistance is required, the services of a competent professional person should be sought.

Advantage Media Group is proud to be a part of the Tree Neutral® program. Tree Neutral offsets the number of trees consumed in the production and printing of this book by taking proactive steps such as planting trees in direct proportion to the number of trees used to print books. To learn more about Tree Neutral, please visit **www.treeneutral.com.** To learn more about Advantage's commitment to being a responsible steward of the environment, please visit **www.advantagefamily.com/green**

Advantage Media Group is a publisher of business, self-improvement, and professional development books and online learning. We help entrepreneurs, business leaders, and professionals share their Stories, Passion, and Knowledge to help others Learn & Grow. Do you have a manuscript or book idea that you would like us to consider for publishing? Please visit **advantagefamily.com** or call **1.866.775.1696.**

TABLE OF CONTENTS

INTRODUCTION
I MIGHT SAVE YOUR REPUTATION

The goal of this book is not to make you a pest control expert but to share fundamental best practices that will help you choose the best pest control company to protect your business and your reputation. Some of you who have large properties or sensitive accounts must make important decisions involving tens of thousands of dollars. Some of you have major pest infestations that need to be eradicated, and some of you are simply looking for the best company to perform regular services or prevent future problems. Whatever your role or situation, selecting the right pest control company is not a decision that should be taken lightly.

So what can good pest management do for you?

- **Save you money.** By avoiding large, costly pest infestations, you won't have to continually throw money at them. You can prevent problems before they start, thereby keeping customers and tenants happy.
- **Save you time.** By reducing the number of times you have to schedule a company to come out and take care of your problem, you'll spend less time looking through service records trying to find what's been done and what hasn't been done and less time tracking what is going on at the property.

1

- **Save you complaints.** And that can just make life more enjoyable. Complaints can come not just from customers but also from demoralized employees. Nobody wants to fear taking bed bugs or cockroaches home from work.
- **Increase performance or profitability.** In apartments and hotels, for example, that comes as a result of increasing the occupancy rate and decreasing turnover of tenants.
- **Prevent devastating negative publicity.** Online reviews, news-media reports, or just word of mouth can drive away customers and are distressing to employees.
- **Make you look good.** When you are the manager or employee responsible for choosing a pest control company that solves problems, saves money, and saves time and hassles, you look like a genius.
- **Save you from lawsuits.** Like it or not, we live in a litigious society where pest control problems can lead to lawsuits involving food poisoning, food contaminations, bed bugs and cockroaches, and neglect of property.
- **Make life better for tenants, customers, managers, and owners.** Quality pest control improves your life and the lives of your tenants, guests, or clients. It protects food, health, and property. It gives you peace of mind that your business and property are well protected.

I'll be using some scientific and industry terminology in this book, but only enough so that you will be familiar with the jargon you might hear from contractors. You'll learn why I obtained a professional credential that involves continuing education and research. You'll find that I love what I do. Giving people a pest-free home, a pest-free night's sleep, a pest-free vacation, a pest-free meal, gives them a better life. Whatever business you're in, you're serving people by having better pest control, and you're improving their lives. And that has to feel good.

CHAPTER 1

WHAT GOOD PEST MANAGEMENT LOOKS LIKE

C hoosing a good pest control company is hard if all you have to go on is what a salesperson tells you. They all say that they're able to handle your problems. So how can you discern which company will offer you an exceptional, hassle-free experience? Look for the following as you compare your options.

EXCELLENT CUSTOMER SERVICE

A good pest control company is easy to work with. It should respond quickly to your emails or messages. It should have a well-staffed office so that whoever answers the phone is friendly, helpful, and knowledgeable enough to answer basic pest control questions and explain how and why the company does what it does.

I suggest calling the company's main number and not the salesperson. Talk to the office staff, and ask them some questions. Find out what it's going to be like to work with them. Are they pleasant? Are they knowledgeable?

QUESTIONS FOR DETERMINING A PEST CONTROL COMPANY'S LEVEL OF CUSTOMER SERVICE

- **How quickly can you respond to emergencies?** This is difficult because good companies are in high demand, which means they're also busy. However, they should be able to respond to emergencies within a reasonable amount of time.

- **How do you treat for a specific pest?** You are trying to find out if the office can easily explain their protocol. You want to know how they will be treating and not just assume the company will do it correctly.

- **What happens when I have a complaint?** Find out if they have a process on how to handle complaints, and get a glimpse of what will happen if mistakes are made.

PROFESSIONAL DRESS AND EQUIPMENT

Good pest management *looks* professional. You have worked hard to build your company and your reputation. Who do you want showing up on your property? Do you want a truck in front of your building with huge pictures of bed bugs all over the sides of it, flashing lights on top, and an oil leak? Or do you want a clean, professional truck showing up with organized equipment and a technician wearing ironed clothes and a clean, tucked-in shirt? It makes you feel good about the company you've hired if that's what its technicians look like. And the people at your location(s) are more apt to listen to somebody who looks sharp and drives a clean truck.

Appearance says a lot about how any professional carries out his or her job. When employees who work in the public realm don't care about their appearance, there's a higher chance they don't truly care about their work. On the other hand, employees who take care of their appearance and equipment are more likely to care about the quality of their work and the level of service they provide.

How do you find out about a company's appearance? The salesperson you talk to will generally look pretty nice. However, before you begin service, ask for a technician to stop by to

Professional conducting an inspection.

make sure he or she is clean and professional looking. What does the truck look like? How does the company present itself?

CREDENTIALS

Beyond appearance, be concerned about credentials. Credentials don't guarantee that the pest control company is great, but they are a good indicator. Why? Credentials show that the company cares about and is involved in the industry. They point to a company that is staying on top of best pest management practices and that cares about education and learning. On the other hand, a lack of qualifications can indicate that a company is just out to make a buck. That's not always the case, but be aware that it can be an indicator.

What credentials should a pest control company have? A good place to start is to find out if they are members of the National Pest Management Association (NPMA), a nonprofit trade association that represents the interests of the professional pest management industry. The NPMA provides education through seminars and conferences, an online information exchange forum, and weekly news updates on research and new products.

It is also good to find out if the company has attained the QualityPro mark of excellence in pest management. The NPMA grants QualityPro certification to companies that have the highest standards in hiring, training, insurance, contracting, marketing, and communication. Additionally, 10 percent of QualityPro companies are audited every year to make sure that they are living up to these standards.

Finally, find out if the company is a member of the Entomological Society of America (ESA). The ESA offers journals, educational opportunities for pest management professionals, and certifications for Associate Certified Entomologist (ACE) and Board Certified Entomologist (BCE). These certifications show that a pest professional can accurately identify insects, recognize evidence of their presence, and apply best practices for prevention and eradication.

QUESTIONS FOR DETERMINING A PEST CONTROL COMPANY'S CREDENTIALS

- **What are your credentials?**
- **Is the company a member of the NPMA?**
- **Is the company QualityPro certified?**
- **Are your professionals members of the ESA?**
- **Do you have an ACE or BCE entomologist on staff?**
- **Are you a member of the local pest control associations?**

KNOWLEDGE AND EXPERIENCE

A good pest management company doesn't only have the credentials; it has the knowledge needed to solve your problem. Service quality comes down to what the service professionals know. Amateur treatments are going to lead to amateur results. Your business does not need or want amateurs. It needs knowledgeable professionals because when you hire a pest control company, you're hiring it for what it knows. That's what you're pay for. *Knowledge* is what we have of greatest value. The more we know, the more valuable we are, and the better we can handle your pest control problems.

Any person with a truck and a sprayer can start a pest control company, but it's not necessarily the person you're going to want on your property. Anyone can spray baseboards with a product, but a pest control *professional* is someone who continuously studies industry best practices. These professionals are equipped to solve problems, and they carry the tools and equipment needed to be successful. They are trained to see things that others miss, and therefore, they will solve problems that others cannot.

Pest control professionals don't just carry around one product in a big tank all day. They need different products for large flies, mice, cockroaches, stored-product pests (a group of insects that includes Indian meal moths, cigarette beetles, and confused flour beetles), small flies, ants, spiders, and anything else they might run up against. Every product and treatment method has its advantages and disadvantages—and they need to understand those and be able to choose the appropriate methods. If the technicians only carry a backpack sprayer and a hand canister sprayer, they are not going to do the job adequately. They should be carrying a variety of other tools like dusters, bait guns, vacuums, granular spreaders, ultra-low-volume (ULV) applicators, traps, exclusion materials, foamers, glue boards, baits, screwdrivers, spatulas, flushing products, and other tools.

One of my favorite things about this industry is that there is so much to study and learn. You can never know everything, and there is always new research, new insights, best practices, and new products to help us do our best—and that's exciting to me. Just because someone has been doing pest control for thirty years does not mean he or she knows everything. There's always more to learn, always ways to improve.

Professional service vehicle with proper equipment.

I'm really quite blessed that I found this industry. I can't think of anything more interesting to be doing with my life. We get to inspect and identify insects and conducive environments, prescribe chemical and nonchemical treatments, and follow up and communicate the results. When things don't go as planned, we get to problem solve and dig deeper to find out what's happening. I get to be a doctor, an investigator, and sometimes a therapist all in the same day, and it's really quite exciting.

QUESTIONS FOR DETERMINING IF A PEST CONTROL COMPANY'S EMPLOYEES ARE KNOWLEDGEABLE AND EXPERIENCED

- **What is the training program for technicians?**
- **Is ongoing training required?** If yes, how often and for how long? Who teaches the sessions? What do they cover?
- **How do you treat for specific pests, such as mice or ants?**
- **Can you explain the science and reasoning behind the treatments you use?**
- **What products do you use for specific insects?** Is it the same product or just a few products? (It's not always necessary to use a different product for every type of insect—but it's not a good sign if they use a single product for everything.)

A DESIRE TO HELP

A pest control company should have a strong desire to help solve your problems, make your life easier, and prevent future infestations, thereby saving you money. If there's not a strong motivation to help, things are going to be missed, and results are going to be lackluster.

There are six signs to look for that demonstrate that the people at a company desire to help:

1. They generally offer long warranties so that you know problems are going to get resolved and that they are

going to stand behind their work. They do not void their warranties.

2. They educate you, your employees, and your tenants (if applicable) about pest control problems and what you can do to help prevent pests.

3. They want to solve the problem, so they thoroughly inspect the property and the pest vulnerable areas. They search the pest vulnerable areas (PVAs) for sanitation issues, and they either fix them or help you find a way to fix them.

4. They find ways to match their services with your needs. They build custom programs for you because not everybody's needs are the same.

5. They communicate and document results. They want you to know how things are going.

6. They talk to you about your responsibilities and how you can help.

When a company wants to help, service is about you and not about the company. That being said, I don't want you to misunderstand and think that pest control companies should do just what you say or that you should be designing the program and telling them what to do. The pest management company should be designing the program, and you'd be wise to listen to its recommendations. However, the program should be designed around you. It should not be a one size-fits-all service. You do not want to hear, "This is all we do for apartments" or, "This is all we do for restaurants." If you historically have dealt with flies and mice, then those two pests should have a prominent place in the pest management program. The company should make sure that the services are built to address the specific pests that trouble your property.

PEST VULNERABLE AREAS

Pest vulnerable areas, sometimes called PVAs, are simply areas that are at high risk for pest activity. These may be obvious, like a dumpster or trash room, or something that's found only after an expert inspection. In either case, PVAs should be identified by the pest management company and written into a commercial account's pest program so they can be inspected frequently.

At a large food-processing plant, for example, we can't inspect every square inch during every service call. But if we have identified the twenty-one most likely PVAs, we can inspect those twenty-one areas every time we service the account and other areas on a less-frequent schedule. Pest control professionals should be trained and adept at identifying the PVAs during initial inspections, whether they are at residential, commercial, apartment, or hospital facilities. The better that they can do this, the faster they're going to be able to locate and eliminate pest activity.

When you have a problem with flies, you don't need the baseboards sprayed. You need insect light traps (ILTs), fly bait, fly stickers, and inspections to find out where the flies are coming from. So how do we build these customized programs and know what problems exist or could arise in the future? We start with a detailed audit of the account, then we complete an inspection of the facility, and finally, we interview the person responsible for pest control in order to find out what problems the account has had.

A good pest management company that wants to help is also going to talk to you about necessary nonchemical controls, some of which will be your responsibility. I know this is not what all managers think they want, but they should. Why? Because pest management companies can't do it all. I know that's shocking, but this is a partnership, a team. You want a company that will tell you the things you can do to help solve or prevent problems. Not addressing these issues is equivalent to putting a Band-Aid on the issue—just covering up the problem is not the same as actually solving it.

The list of preventative, nonchemical controls can include: getting access to adjacent units or other areas of a facility; trimming vegetation back from a building; fixing water leaks or removing standing water outside; repairing cracks and structural issues; pest-proofing doors and sealing windows; talking to tenants and giving notices when they're needed; and cleaning up spilled food, dirty drains, and clutter next to the facility. All of these things can have an effect on pest activity, so the pest control company should be talking with you about these issues and finding ways to fix them.

Before you hire any company, you should meet with the people and ask them questions to find out if they're a good fit for your needs and if they have that desire to help. The more in-depth this discussion is, the better your selection is going to be and the better the company will be able to understand and meet your needs. Those needs must be understood if a program is going to be successful.

QUESTIONS FOR DETERMINING A PEST CONTROL COMPANY'S DESIRE TO HELP

- **What will our pest control program look like?**
- **How will you build it to meet our needs?**

- **What are your goals for our property?** (They should mention their desire to improve your life and well-being.)
- **How does your company report pest-conducive conditions?**

PROACTIVE PROGRAMS

Good pest management involves designing proactive programs. We need to be playing offense and not just defense. We need to be looking at tracking and fixing potential problems. So before you select a pest control company, look at its program and ask yourself—and possibly the pest control company—does this program prevent problems before they start? Does this program actually prevent bed bugs, cockroaches, mice, spiders, or flies? Or does it simply treat them?

If the program that a company puts together for an apartment complex is only saying, "We're going to treat up to three complaints every time we come out," that's not preventing anything. Are there inspections and monitors built into the plan that are going to catch problems when they crop up? If so, it is a proactive program. Not all facilities need a lot of monitors, but all commercial accounts should at least have regular inspections, otherwise it will be difficult to find out what pest problems exist. Are you just going to wait for complaints from employees and tenants? Are you waiting for an inspection from the health department? If so, you may be putting your business at risk.

Tenants in particular are reluctant to report pest problems. They may be embarrassed that they have cockroaches, bed bugs, or flies and afraid they will be evicted or have to pay for the treatment. The tenants with the worst infestations may be the least likely to report. Pest control companies that don't inspect and get to the source of the problem in order to solve it can make a lot of money because the problems never get resolved. The company just keeps treating and treating while the problems spread and get worse, costing more and more money. I sincerely hope this is not what most pest control companies do, but if they're giving a short, thirty-day warranty and they're simply spraying and decreasing the insect population, things may seem to improve—but only for a short time. Then, two months later, if it's an apartment complex for example, one tenant calls in and says, "I'm having an issue with cockroaches," and then the next tenant calls. If we're constantly just treating the units that call in and we're never finding the sources of problems or treating the whole thing, we're never going to solve anything. Sadly, we often see properties spending large amounts of money on pest elimination but never achieving long-term results with their pest problems.

We've taken apartment complexes that were spending $90,000 a year on bed-bug control and reduced their costs to around $12,000 a year for mostly preventative services. It would be great for us to be making $90,000 on an account like that year after year after year, but it's probably not sustainable. And even if it was, it's not the right thing to do. A focus on prevention saves businesses money and ensures that efforts will yield lasting results.

WHAT DOES A PROACTIVE PROGRAM LOOK LIKE?

Inspections allow us to look for and locate problems and potential problems. They help us know what's happening on the property. They help us locate pest infestations and pest-conducive conditions.

Monitors are our eyes and ears on the property 24/7. A variety of different monitors exist for mice, flies, spiders, crawling insects, stored-product pests, or termites. Monitors such as light traps, bait stations, and glue boards can tell us if our treatments are working and if different treatments are needed.

Preventative treatments help us prevent infestations before they happen or before they cause major problems. Because flies become more active in the summer and fall, we can perform preventive treatments in the late spring or early summer. These may involve chemicals or physical barriers.

QUESTIONS FOR DETERMINING IF A PEST CONTROL COMPANY IS PROACTIVE

- **Are proactive inspections part of your pest control programs?**
- **Do your programs include monitoring for pest activity?**
- **If so, what do you monitor? Where do you normally place monitors, and how many monitors do you normally use?**

INTEGRATED PRACTICES

A good pest management company practices integrated pest management, or IPM, which means performing detailed inspections, identifying the pests involved, employing multiple treatment methods including chemical and nonchemical-control methods, and following up and communicating those results. Everyone says that they are doing IPM, but it's important to know what IPM is and if that is actually what's being done.

Inspections are the first part of IPM, the most crucial to a successful pest management program, and probably the most-often skipped or hurried step by pest control professionals. Some of the service time for your business should be dedicated to inspections. Professionals should inspect the interior and exterior PVAs regularly using a flashlight, screwdriver, flushing pesticide, glue boards, a spatula, a magnifying hand lens, vials to collect specimens, a good camera, and something to take notes on. During the inspection,

Professional performing a thorough inspection.

the professional should be looking for insect activity, pest harbor-ages (locations where pests seek shelter), conditions that may be conducive to pests being there, and other areas that are vulnerable to pest issues. A good pest control company is good at inspections.

Effective recommendations and treatment(s) should follow the inspection. Good pest control companies provide reports on the deficiencies and the pests that are found in these inspections. They should also clearly communicate who will fix the issues and when and how the problems will be resolved.

The following table includes a list of the most common types of nonchemical and chemical-control methods. Your pest control company should be familiar with these options.

NONCHEMICAL CONTROL	DESCRIPTION
Harborage Removal	removing areas or items that provide shelter for pests
Sanitation	removing conditions that provide food and water for pests
Exclusion	sealing up or eliminating cracks, holes, and entry points that allow pests to enter the building
Trapping	placing devices that physically catch the pests
Heat Treatments	eliminating pests by raising the temperature typically to 120–150 degrees for an extended time
Cold Treatments	a method of eliminating pests by lowering the temperature to a lethal range
Vacuuming	simply removing pests with a vacuum—providing immediate reduction in the number of pests

CHEMICAL CONTROL	DESCRIPTION
Exterior Perimeter Treatment	treatment of the exterior of a building to prevent pests from entering the building—generally including treatment around windows and doorways, exterior foundation, and likely insect harborages near the facility
Spot Treatment	a treatment no larger than two square feet made where pests are likely to enter, move over, or harborage
Ultra-Low-Volume (ULV) Treatment	treatment for knocking down flying insects and some crawling insects, where a machine creates tiny droplets of pesticide that can float in the air for several hours
Bait Treatment	baits with active ingredients placed in food or other attractive materials to get the pests to feed on them
Crack and Crevice Treatment	treatments where residual pesticides are placed into cracks and crevices where pests enter, travel, or are likely to harbor
Granular Treatment	dry formulations in granular form where pesticide is attached to clay or nutshells and is often placed around buildings in mulch, lawn, and window wells
Dust Treatment	treatment of a fine pesticidal dust into cracks and crevices, wall voids, behind objects, and other inaccessible places—typically has very long residual, and insects pick up the material easily

Follow-up assessments should take place after the appropriate treatment methods have been put in place to ensure they were effective. The follow-up results should be reported back to you, and if needed, more treatments should be done or scheduled. The company should clearly communicate what was done, how things went, and if any follow-up is necessary.

QUESTIONS FOR DETERMINING IF A PEST CONTROL COMPANY IS REALLY USING IPM

- **How do you approach a normal treatment?** If they omit any of the steps we just covered—inspections, identification, comprehensive treatment, and follow-up—they're not practicing true IPM.
- **What methods do you use to treat different pests?** Find out if they're doing different things or if they're just spraying baseboards.
- **How do you treat for ants (or any specific pest)?** If they jump right into treatment, they're not doing IPM. They should first inspect and find out what type of ant it is, and that will tell them how to treat and how to follow up to make sure it's taken care of.

If you ask those questions, you will have a good idea if the pest management company is really practicing IPM.

DETAILED DOCUMENTATION

How do you know what pest activity is taking place in and around your facility? Through tracking and documentation—critical activities for any commercial account, whether it's a food-processing plant, restaurant, hospital, apartment, or any other type of commercial property. Documentation shows you what is being done, what you are spending your money on, whether there are problems that need to be addressed, what areas are having pest activity, and what areas have potential activity—or what we call conducive conditions. Are there trends of pest activity that need to be addressed? Is what the company said it was going to do being done? The answers to these questions can be found in the documentation, if it exists.

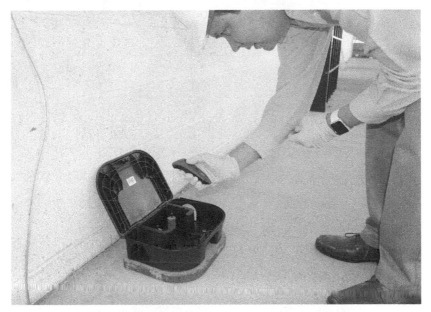

An example of digitally capturing information with a scanning bait station.

Digitally capturing data is important in this day and age. If one of our clients wants to know if bait station sixteen at a property has had activity in the past six months, we can provide that information in twenty seconds if we have the proper documentation. We don't have to thumb through six months of papers or service reports anymore. The information is digitally captured when our technicians scan the barcodes on the devices we've installed at our commercial accounts and record their observations.

Professional is scanning an insect light trap to document the number of insects captured.

The types of devices we install include: rodent bait stations, multi-catch monitors, ILTs, fly bait stations, and stored-product pest monitors. Any time we service the site, we can tell our clients the exact number of pests we've monitored and caught. We can also run trend reports to show what's happened in the past twelve months on any or all devices. We can track exactly how many flies a device has caught every week to show if a fly program is working. If a client needs to know if apartment B16 has been treated for cockroaches, we can pull that up and know immediately, without having to dig through all the service records. I've met many residential and hotel property managers who have kept their own paper records or spreadsheets to track pest activity and create a record of which units have been treated for bed bugs, spiders, and cockroaches. Your pest control company should do all that tracking for you.

QUESTIONS FOR DETERMINING IF A PEST CONTROL
COMPANY KEEPS THOROUGH DOCUMENTATION

- **How do you monitor and track pest activity at the site?**
- **How do you use documentation to inform your treatment and prevention methods?**

CLEAR COMMUNICATION

Beyond documentation on paper and in digital form, pest management professionals should verbally communicate with the manager or contact person every time they're on-site. Before and after the work is completed, the pest management professional should address these four things with the contact at the property:

1. What they found
2. What they did and why
3. Any potential problems that they see
4. Any follow-ups that are needed

QUESTIONS FOR DETERMINING IF A PEST
CONTROL COMPANY COMMUNICATES WELL

- **What types of reports do you provide clients?** (Ask to see examples.)
- **How often do you do an audit of your accounts?**
- **Do you provide clients with trend reports for the year?**
- **How will I know exactly what is going on at the property?**

TIMELY AND CONSISTENT RESULTS

The most important sign of a good pest management company is the end result! Results should be real and should be documented with follow-up inspections and reports. Your pest control company should not just treat and then hope everything goes well; it should be able to provide clear evidence of progress. For example, it might say, "After the first treatment our monitors caught 350 cockroaches, after the second treatment we caught sixty cockroaches, following the third we caught twelve, and most recently our monitors caught zero cockroaches." That's how you know that things are happening—by the evidence that shows they are going back, following up, and inspecting to make sure the job was completed successfully.

Things should be looking better at your facility month after month. That doesn't mean problems aren't going to pop up; however, if the program is well designed, problems are much more likely to be identified and addressed before they become widespread and costly. I should stress here that implementing a good pest management program is difficult if you don't have a proper budget. For example, a monthly pest control budget of sixty-five dollars for a huge grocery store or apartment complex isn't enough to proactively stop problems before they arise. On the other hand, sixty-five dollars a month could possibly be enough to properly prevent and control pests at a much smaller and less complex facility. You and your pest management company should come up with an appropriate budget after considering past history, vulnerable areas, and your tolerance threshold. An office complex has a different threshold than a hospital should have, which we will elaborate on later in the book.

Just remember that proper inspection and prevention is worth the cost. You don't have to go crazy with your pest control budget, but you do have to listen to your pest control company and make sure you've allocated enough money to build a sufficient program.

Simply reacting to problems will generally cost you more money in the end because problems won't be identified fast enough and will spread throughout the facility as a result. Some companies are expensive to hire and do bad work—but still somehow maintain good relationships with their customers. The price tag and the people's likability are not good measures of the time they are spending, the effort they are putting in, and the results they are getting.

QUESTIONS FOR DETERMINING IF A PEST CONTROL COMPANY IS RESULTS DRIVEN

- **How long will it take to resolve the problem?** (You want a rough estimate so you can determine if you're both on the same page. You don't want to be thinking that a problem will take two weeks to resolve while the pest control company thinks it will take six months.)
- **What criteria do you use to determine if the property has been properly treated and the problem is solved?**

WARRANTIES

A good pest management company is accountable. No company is perfect, so mistakes will sometimes be made, but does the company admit them when it does something wrong or try to fix mistakes when possible? You don't want to constantly hear excuses. Some companies will blame everything on the tenants and on the weather and on the products. Everything but them is the problem. Pest control companies should be designing services that, in most cases, work regardless of these factors, no matter how dirty a tenant is, no matter what the weather is.

I've never understood why some pest control companies offer such short warranties. I suspect part of it is to cover up failures and make more money, but you should be looking for companies that offer long warranties to make sure that the problems are resolved, especially for pests such as mice, cockroaches, and bed bugs.

QUESTIONS FOR DETERMINING IF A PEST CONTROL COMPANY IS ACCOUNTABLE

- **How long is the warranty?**
- **What are your general pest control warranties for things like mice, ants, and wasps?**
- **What happens if treatments are not improving the situation?**
- **What happens if I'm not happy with the quality of your service? Can I get out of my contract?**

WARRANTY GUIDELINES

Situations vary, but the following are some guidelines for warranties that are long enough to show that companies stand behind their work. Particularly with bed bugs, German cockroaches, and termites, you should be getting 100 percent eradication.

Pest control is an investment; therefore, you want to make a good decision when hiring a pest control company. To ensure you're partnering with the right professionals, ask enough questions to ensure that they are knowledgeable, competent, reliable, easy to work with, and—most importantly—focused on solving your problems for good.

PEST	WARRANTY
Bed Bugs	six-month to one-year warranty against live bed bugs after treatments are completed
German Cockroaches	three- to six-month warranty after cockroaches have been eliminated
Termites	one to five years (the longer the better)
Ants	one to six months (there are many different ant species, and this may vary widely depending on species and situation)
Carpenter Ants	three to six months
Rodents	depends on the situation and the pest pressure (the company should stand behind its work and eliminate the problem, eliminate conducive conditions, exclude from building, and prevent and monitor for future activity)
Flies	should see significant reduction—determined by monitoring
Stored-Product Pests	varies widely depending on the situation and facility (make sure the company is following best-management practices and inspecting, trapping, eliminating, or helping identify conducive conditions and monitoring)

GUIDE TO INSECT LIGHT TRAPS (ILTS)

What are they? Insect light traps, or ILTs, are electrical devices containing several light bulbs to attract insects and some kind of trap, typically glue boards. Some also contain a bait to attract flying insects, such as flies, moths, and fruit flies. ILTs come in different sizes. They are hung on walls. Some are discreet; others are obvious insect traps.

Placement: When properly placed, ILTs can be very effective at decreasing the number of flying insects and monitoring their activity. But a lot of companies get it wrong. Hiring someone who doesn't know how to properly install a light trap is a good way to waste money. You should know the basics so you can make sure your ILTs are properly installed, maintained, and documented at your facility.

Your pest management professional should start with a site map of your facility and try to determine how the flying insects would move from room to room. After identifying possible locations for ILTs on the map, he or she should walk through the facility with you to determine if those sites are going to work well.

The ILTs should be installed two to five feet high for flies. (That's the ideal flying height for flies.) The first line of defense needs to be fifteen to twenty-five feet from the entrances. If closer than that, the insects are often going to fly right past the fly light, and you're not going to get nearly the results that you want.

To avoid attracting insects from the outside, the ILTs should not be visible through windows or doorways. They should not be in areas with a lot of competing light. Darker, windowless areas are better, especially areas where insects will naturally be going, like trash rooms. Placing ILTs in skinny hallways can help their effectiveness because the flies don't have as much room to avoid them. In a food plant, they should be placed in areas leading to production so you can catch as many flies as you can before the production area. ILTs should never be placed above food-prep areas. Placing them away from food prep draws the flying insects away from the food.

Maintenance: Properly maintaining ILTs is important to ensure that they stay effective for many years. The light bulbs should be replaced at least annually. Lights should be cleaned regularly of debris and dead insects. Glue boards must be replaced at least monthly or more frequently in dusty areas or areas of high insect activity.

Documentation: ILT catches should be documented at least monthly and maybe more often depending on the account. Pest management professionals should document what insects were caught and how many were caught. We should collect this data at least monthly and periodically run trend reports to look for patterns, analyze the results of treatment, and see what else needs to be done.

I also meet new clients who do not keep any record of past activity or previous treatment efforts. Oftentimes, we'll find bed bugs at an apartment complex, and I'll say, "I saw some dust and insecticide residue, but I still found live bed bugs. Has this area been treated before?" And the answer I get nine times out of ten is, "I don't know." So I ask, "Is there any way to find out?" And they say, "Well, not really. I can dig through all my records for the last six months, but it's probably not worth it, so just go ahead and treat the apartment." They're wasting money. You can save time and money by hiring professionals who conscientiously track these things for you.

CHAPTER 2
PEST MANAGEMENT BY INDUSTRY

You read about good pest control in the previous chapter, but what does really bad pest control look like? Before they called us, the owners of a large apartment complex had already spent $200,000 in one year with three different pest control companies trying to solve its pest problems. Unfortunately for them, despite the money they had spent, their more than three hundred units were still infested. The other companies' treatments were not working; the pests were driving out tenants, and the 75 percent vacancy rate was cutting deeply into revenue.

The owners finally called us at Thorn Pest Solutions, and we spent a couple of weeks just doing inspections. Once the inspections were complete, we determined that 87 percent of the units were infested with German cockroaches. In six months, we were able to remove all cockroaches from all but 3 percent of the units, which had only minor problems, and we will continue to monitor and treat until they are all eliminated. Last time I checked, the occupancy rate was up to 95 percent, the manager was no longer receiving complaints, and tenants were enjoying healthier homes. Some residents who thought they would never be rid of cockroaches were in tears thanking us. We had people baking us cakes!

A good pest management program is important for all industries. But different types of businesses have different challenges and require customized solutions, which this chapter will describe.

MULTIFAMILY HOUSING

The difficulty in apartment complexes starts with having a lot of people living in close proximity. If one of the neighbors is not doing the dishes, not taking out trash, or hoarding, sanitation becomes an issue. Multifamily housing has a lot of doors, windows, and other openings where pests can enter. Plumbing and utility lines allow pests to

Hoarding contributes to pest problems.

move from apartment to apartment. Turnover of residents creates opportunities for new people to bring in new pest problems. And management may not even know because tenants are reluctant to report pest problems for various reasons that we will discuss later.

With all those challenges, we want to build a good program to detect, prevent, and control pests. The most important thing missing from a lot of pest control programs in multifamily housing is regular and thorough pest inspections for bed bugs, German cockroaches, and mice in all apartments. In addition to the lack of inspections, there's often very little monitoring or follow-up. I also see a lack of clear documentation concerning what units are being treated and what is being found and done in each. This documentation is really important, and it should be done digitally, where it's uploaded into a computer and stored for easy, reliable retrieval. Which is much more convenient than a mere paper record.

INSPECTION

The more problems that are documented in an apartment complex, the more regularly the apartments should be inspected. And these should be proactive inspections, which entail looking for issues, not just responding to complaints. In buildings that have had chronic and severe problems, we might have to inspect every apartment monthly, but usually four inspections a year is enough to stay on top of things and catch new infestations early. If you have fewer problems, you can back off and save some money by having inspections in all apartments once or twice a year. But no matter the history, I strongly suggest all apartment complexes should be inspected at least once a year.

Your pest management company should also be inspecting the exterior of the buildings regularly, looking for pest-entry points, vegetation that's too close, heavy activity of mice, as well as anything that needs to be moved off the building like wood, tires, equipment, etc.—generally speaking, anything that's going to cause conducive conditions for pests. Exterior conditions should be documented in inspection reports. Identifying entry points where pests are getting in and sealing them up is critical. A metal mesh that doesn't rust generally works much better than, say, a foam sealant. Good door sweeps and tight-fitting screens also help.

TREATMENT

Preventive treatments should be done inside and outside. Sanitation is huge in preventing problems with German cockroaches, mice, and ants, so it's important to help tenants clean up and remove harborages. A great way to monitor and evaluate progress with cockroaches and ants is to use sticky traps in their high-runway paths. Multiple-catch monitors can also be placed for mice. These are follow-up mechanisms to make sure that whatever pests you're treating—whether ants, cockroaches, mice, or bed bugs—are eradicated and not infesting other units.

MONITORS

A monitor is any device that either traps a pest or tells us if pests are present. Monitors that capture pests can help us identify them and determine if treatments are necessary. They can also tell us if treatments already underway are effective, help us determine the severity of an infestation, and provide us with information that we can use to identify trends.

Every imaginable pest problem can affect multifamily housing, but bed bugs, cockroaches, ants, and rodents are probably among the worst. There are protocols for attacking each, and these are explained in more detail in upcoming chapters. Apartment managers need to be involved and talking with the pest control professionals on-site about what's going on. This is a brief overview of some of the options that may come up in planning a pest control program designed for individual situations in an apartment complex.

- **Bed bugs:** I strongly recommend inspections, monitors, and preventive treatments to save money long term in multifamily housing. There are some effective dust treatments that can, at best, prevent bed bugs, and at a minimum, knock them down, contain numbers, and keep them from spreading from unit to unit. If live bed bugs are found, treatments depend on the severity, the nature of the property, how much stuff is in the apartment, and what the tenants agree to.
- **Cockroaches:** Chemical treatments or baiting treatments are fairly standard, but the follow-up, inspections, and monitoring is often lacking. Make sure all these steps are taken. The number of cockroaches caught in a monitor should be counted and documented in order to monitor progress.
- **Ants:** Professionals should do inspections, identify the ant species, determine what treatment methods are best, set baits, and treat the colonies directly. Just spraying baseboards will often not give you the results you want.
- **Rodents:** The protocol must include inspection, exclusion (keeping them out of the buildings), sanitation, and use of exterior bait stations and interior traps. Just placing traps along the walls is a waste of time. Traps and devices should be placed where rodents are living and moving, places found by understanding rodent behavior and inspecting thoroughly.

DOCUMENTATION AND EDUCATION

Before we do any of this work at an apartment complex, we collect information on how the buildings and units are configured and numbered. Then we enter those building names and apartment numbers into our computer system. That way, when our professional shows up, he or she has the ability to store a digital record of what was observed and done in each unit, and we can retrieve the information easily. We need to document what was caught in the monitors, as well as any deficiencies in sanitation and exclusion problems. Then we can run regular trend reports showing any patterns of problem areas.

Education is also important to a good pest management program for multifamily housing. If the staff on-site learns how to recognize signs of the major problem pests, they can be the eyes and ears on the property. They don't have to become experts, but the more they know about the pest control program the better they can communicate with tenants about it and the better everyone can help control these pests. Educating tenants on key pests and sanitation is also very helpful.

RESTAURANTS

Getting daily food deliveries opens restaurants up to lots of new pest issues. The food and the moist environment that comes from constantly washing floors and dishes are conducive to pests. If a company is just going through and spraying the baseboards, that product is simply being washed away. Meanwhile, food is being trapped in machinery, and food and grease are falling behind items in a busy kitchen and not being cleaned up. Also, the warmth of the restaurant and those food smells really can attract some pest issues. As a result, restaurants pose unique challenges when it comes to pest management.

Some restaurants have a great pest control program that prevents problems without contaminating any food. However, too

Thorn Pest Solutions
Ag. Lic. 4000-436
484 W 220 S
Pleasant Grove, UT 84062
800-626-1156

Service Inspection Report

ORDER #: 52795

WORK DATE: 3/24/16

BILL-TO	110862	LOCATION	110862		

Time In: 3/24/16 1:37 PM
Time Out: 3/24/16 3:13 PM
Customer Signature

Customer is unavailable to sign
Technician Signature

Eric Nelson

Eric Nelson
License #:

Purchase Order	Terms	Service Description	Quantity
None	COD	Inspection	18.00

GENERAL COMMENTS / INSTRUCTIONS
None Noted.

CONDITIONS / OBSERVATIONS	Reported	Severity	Responsibility	Reviewed
None Noted.				

PRODUCTS APPLICATION SUMMARY
None Noted.

PEST ACTIVITY	# Areas	# Devices	Pest Totals
	1	0	Evidence
Ants	2	0	8
Cockroaches	11	0	56
Mice	1	0	Evidence
Spiders	3	0	1

DEVICE INSPECTION SUMMARY

AREA COMMENTS
Unit 1: Didn't see any roaches or BB. Resident says they have seen spiders and "little black beetles". Unable to find any to show me.
Unit 10: Key didn't work
Unit 11: Old roach droppings, no live roaches.
Unit 1A: Saw two roaches while ... a few MW in the spray ...
Unit 2: Didn't see anything.
Unit 2A: Resident has constant issues with ants. Didn't see any roaches.
Unit 3: A lot of spider webs in home. And roach droppings.
Unit 3A: Didn't see any other pest activity. But did see some fecal stains under the sink that was from roaches in the past.
Unit 4A: Saw four roaches under the sink.
Unit 5: Have roaches.
Unit 5A: A lot of fecal droppings from roaches under the sinks. About 30 live roaches behind the fridge.
Unit 6: Old roach evidence, nothing new.

INSPECTION DETAIL

Area	Time	Type	Status	Pest Findings
Unit 1	2:20:23 PM	Area	No Activity	
Unit 10	2:53:30 PM	Area	No Activity	
Unit 11	2:56:15 PM	Area	Activity	- Fecal Stains
Unit 12	2:59:28 PM	Area	Activity	Mice - Rodent Droppings
Unit 1A	1:49:53 PM	Area	Activity	Ants - 3 S Cockroaches - 2 S
Unit 2	2:24:28 PM	Area	No Activity	
Unit 2A	1:53:40 PM	Area	Activity	Ants - 5 S
Unit 3	2:30:07 PM	Area	Activity	Cockroaches - Fecal Stains Spiders - Cast Skins or Egg Sacks

Multi-Family Housing Inspection Report.

many restaurants merely have a company that they call when they see an issue or a company that comes to spray the baseboards once a month or once a quarter.

INSPECTION

A good program starts with regular inspections for sanitation and food safety. A pest control company's reports should include any notations during the inspection regarding food not being properly handled or stored. Ideally, restaurants will leave an aisle of sixteen to eighteen inches along the walls to facilitate inspection.

The inspector should look for pest activity inside, outside, and in a random sample of the delivery trucks. The restaurant staff should always be inspecting the deliveries and trucks—but it's good to get the pest control company involved because we are trained to know what to look for and to do it without offending the suppliers.

Extreme sanitation issues in restaurants often cause pest infestations.

TREATMENT

The best treatments in a restaurant are sanitizing properly, cleaning up food and grease, keeping drains clean, and sealing up pest entries. Chemical treatments along baseboards often get cleaned or wiped away, so the pest control company needs to do more than just spray the baseboards. Instead, the technicians should treat cracks and crevices and voids inside the walls where insects harbor. Plus, they need to be using different materials and treatments than you get from the companies that only know how to spray the baseboards.

Flies are often a problem, so the more fly-treatment methods used, the better. Besides keeping them out and eliminating their food sources, treatments include baits, stickers, glue sticks, and light traps. Keeping outdoor trash bins clean and further away from the building also helps.

MONITOR AND EVALUATE

As with any good program, we need to monitor and evaluate the problems in a restaurant. Light traps allow us to track the numbers of flies caught every single month, run trend reports, and determine if treatments are being effective. The lights should be placed about two to five feet high and must not be over a food-preparation area. They can't be visible from outside, or they would attract flies to come inside. Glue boards inside the traps should be changed at least monthly, and the bulbs should be changed about once a year to make sure they continue to be effective.

For rodents, exterior bait stations help us find out what's happening on the outside so that we can keep populations low—while inside, we place multiple traps to catch anything that comes in or monitor any activity. And then we need to meet with the manager regularly to discuss how the program is working and

Device Trend Report

Legend

Inspected	Skipped	Pest Activity
Replaced	Removed	Pest Captured

ALL DATES

Device Types: All

Device	2015 1/12	2015 2/9	2015 3/9	2015 4/13	2015 5/6	2015 5/7	2015 5/8	2015 5/9	2015 5/10	2015 5/12
1										
2										
3										
4										
5										
6										
8			FR - 5	FR - 60 HO - 10		HO - 45	HO - 4	FR - 10 HO - 1	FR - 3	FR - 1
9				IN - 4		IN - 2	IN - 1			
10					IN - 1		IN - 6			
11				MI - 2	RA - 1	RA - 1	MI - 1		MI - 2	
12			MI - 1	MI - 1						
13					RA - 1		MI - 1			
14	MI - 1								MI - 1	MI - 1
15	MI - 1	MI - 1				RA - 2			MI - 1	
16	MI - 1					RA - 1		MI - 1		MI - 3
17			FR - 10 HO - 5	FR - 20 HO - 2	FR - 50 HO - 8	FR - 70	FR - 24		HO - 6	HO - 4

Device Trend Report- HO- House Fly, MI- Mice, RA- Rodents, IN- Indian Meal Moth.

provide service reports and findings from the inspections and monitoring.

EDUCATION

Having a good pest control company train a restaurant staff can help them understand how sanitation and proper storage affect pest control and how their processes and behaviors can help prevent and treat problems.

FOOD-PROCESSING PLANTS

Food-processing plants not only deal with food, but they also receive regular deliveries that can introduce new pests. Warmth and food smells are wafting from the building and attracting pests. Doors are propped open by workers wanting a cool breeze. Machinery is often difficult to clean properly. Many plants are open twenty-four hours a day, leaving no downtime for the pest control company to work. The buildings themselves may be poorly sealed and in locations conducive to pests.

Food-processing plants need to have a very, very low tolerance for pests, so an excellent written program is critically important. The pest control company should be a partner that works with management to keep the food free of contamination. The company should inspect,

A professional inspecting a food-processing plant.

provide treatment options, monitor and follow up on problems, provide detailed and clear documentation, and educate the clients on pest-conducive conditions and how they can improve their efforts to keep pests at bay.

INSPECTION

Inspections should be done constantly. The pest management company should be checking things like sanitation and food spillage, as well as making sure the drains are not harboring pests. They should be looking for any pest evidence, including droppings or sheds or dead insects. Inspection aisles should provide eighteen to twenty-four inches of clearance from the walls.

The pest control professionals should look for sanitation issues, structural deficiencies, and anything else that would create pest-conducive environments. They should also inspect delivery trucks and the exterior of the plant, making sure there's no vegetation or debris stacked up against the building that could give shelter to pests.

TREATMENT

Treatments must be well thought out so that food does not become contaminated. Nonchemical-control methods are used the most; these include killing pests with high heat or cold, vacuuming them up, trapping them, sealing them out, and ruining their habitats. When chemicals are needed, it should be as a spot treatment—and the guidelines on the labels should be followed strictly.

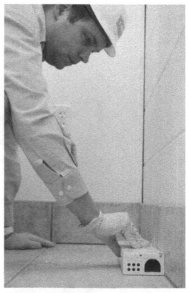

Multiple-catch mousetraps can assist in monitoring rodent activity.

MONITOR AND EVALUATE

Monitoring and evaluation are important to find the pests and evaluate the success of the treatment. For stored-product pests, there are different pheromone traps that can be very helpful in monitoring activity. Rodent monitors, exterior bait stations, ILTs, and insect monitors should also be used in these food facilities to keep an eye on pest issues twenty-four hours a day. Food-processing plants can have many kinds of intruders, but the most common are stored-product pests, mice, cockroaches, and flies.

US food-processing plants have to comply with a relatively new federal law, the Food Safety Modernization Act. According to the Centers for Disease Control and Prevention, about forty-eight million Americans, or one in every six, get sick, and 128,000 are hospitalized from food-borne diseases every year in the United States. To help combat this, the Food Safety Modernization Act was signed into law in 2011. It focuses on preventing food contamination.

DOCUMENTATION

Documentation is crucial, especially when the plant is being audited to make sure it is meeting best-practice standards for food safety and pest management. Inspection reports must show any deficiencies found. Regular reports on all the monitors—including weekly or monthly trends—have to be used to plan treatments. Then come service reports of what's being done and how and where products are being applied.

The pest control company should review your program with you at least once a year. It should also offer to provide some education, quarterly or at least once a year, for the plant staff on what's being done, so they know you are doing everything you can to control pests at the facility.

You and your pest management company should be familiar with the Food Safety Modernization Act and best practices to help keep you in compliance. The requirements include:

- **a written preventive pest management plan**
- **identifying pest vulnerable areas (PVAs) and having preventive controls in place for these risk areas.**
  - Each plant is unique, but some common examples of PVAs are trash rooms, dumpsters, bay doors, food-storage areas, machinery, delivery trucks, break rooms, loading docks, and processing areas. Preventive controls in such areas must be written into the pest management plan.
- **written pest thresholds, with action steps**
  - This means you need to identify potential pests at the food facility and have an action plan that kicks in when a preset threshold is reached. Inspections, monitors, and a pest-sighting log should be a part of the pest-prevention program to determine if these preset thresholds are reached.
- **clear documentation of deficiencies found at the facility**
  - Corrective actions must also be documented, along with who is responsible to correct them.
- **clear documentation for every aspect of pest control**

> ▫ Service reports must be kept on-site, and reports must
> be kept regarding devices or monitors, what's being
> caught and when, along with sanitation reports,
> trend reports, material safety data sheets and labels
> for chemicals used at the facility, pest management
> professional licenses, site diagrams of the facility with
> the location of pest management devices, and pest-
> sighting logs.

For example, in a food facility—where you don't want even a
single mouse—the plan would specify that the mouse threshold is
one. If a mouse or evidence of a mouse is found inside the facility,
an action plan might stipulate that the following five steps should
be followed:

1. Inspect interior and exterior in proximity to where
    evidence of the mouse was found.
2. Identify possible entry points and conducive conditions.
3. Eliminate entry points and conducive conditions.
4. Place traps in likely mouse runways, minimum of eight.
5. Check traps at least weekly.

Such written plans need to be worked up for each pest that we
think is possible at the facility. And if we find that a loading-dock
bay door is allowing pest entry and needs to be fixed or sealed
better, that needs to be documented along with who's in charge of
it. If the plant management corrects the problem, then we need a
document recording that the deficiency has been repaired.

You shouldn't have to do all of this paperwork on your own. It
can be fairly daunting if you don't know what you're doing. So talk
to your pest control company and enlist its help to help put this
together and make sure you're in compliance. If your company is
not able to help—or not knowledgeable in best practices for food

plants—then find a company that can help and provide you with what you need.

HOSPITALS AND HEALTH-CARE FACILITIES

The difficulty is that lots of people are continuously coming in and out of hospitals and health-care facilities, so they can bring in pests carrying diseases that especially threaten patients who may already have weakened immune systems. The big concern is to keep out pests that might cause nosocomial diseases—which are any that the patient didn't come in with but contracted while there.

A hospital can be a complex environment for pest control because there are many doors, lots of deliveries, lots of trash, and lots of laundry. Typically, it's very hard to inspect every square inch of a hospital. A true professional with a well-defined pest control program is needed. Staff should also be well versed on what's going to happen, where the vulnerable areas are, and how to report issues as they see them.

Hospital staff and doctors may not be inclined to take time to listen to a pest control technician, but they need to help implement the program, so having somebody who's educated and professional talk to them helps to earn their respect.

INSPECTION

Companies should be spending a large portion of their time inspecting to gain a clear understanding of the vulnerable areas and how often to search each of them. The company should use pest-sighting logs, which hospital staff can contribute to. It should also inspect the exterior of these facilities. The sources of any pest issue need to be found and treated directly. Sometimes it takes a lot of investigating to follow the trail back to the source—where fruit flies or phorid flies are breeding, for example.

The pests most likely to cause distress or spread diseases in hospitals are ants, flies, bed bugs, mice, and German cockroaches.

TREATMENT

The use of nonchemical control methods is preferred whenever possible; among these methods include sanitation, exclusion, habitat modification, vacuuming, heat treatments, and cold treatments. Responding quickly is crucial because of the very low tolerance for pests in these facilities. So we check those sighting logs regularly and monitor and evaluate results of treatments. A hospital is too large to inspect completely, so using many monitors can really help make sure things aren't being missed. These include exterior bait stations, multiple-catch traps on the inside, insect monitors in PVAs, and ILTs to find out where and how flies are entering the building.

DOCUMENTATION AND EDUCATION

Hospital managers should receive inspection reports, monitor results, trend reports from those monitors, and service reports to document how any issues are being taken care of. Less common, but helpful, is an education program for the staff, where at least once a year everybody's kept up to speed on how to report problems, how they can help, and why pest management is important.

THE HOTEL AND LODGING INDUSTRY

Hotel guests can create sanitation issues and thereby bring in any number of pests. Food gets dropped behind the bed or behind the dressers, creating a food source for various types of pests. In many hotels and motels, rooms are lined up right next to each other, and insects and rodents have runways because of the way electrical wires and plumbing go from room to room. That makes it difficult to control pests—but it has to be done because reputation is critical in the lodging industry. No matter what type of hotel, pests in the rooms cannot be tolerated, because the business reputation will be damaged quickly through either word of mouth or online reviews. So these locations need good, well-designed pest management programs.

INSPECTION

We recommend regular inspections of all rooms, making sure that things aren't getting missed, including bed bugs. Inspections must also include restaurants or breakfast areas, custodial closets, laundry areas, and the garbage areas and dumpsters outside, all of which are pest vulnerable. We should be inspecting the entire exterior of the building, looking for pest-conducive conditions, pest activity, and pest-entry points.

Maintenance staff on-site can help keep pests outside by making sure that doors, pipes, and vents are sealed well. They should make sure that vegetation is trimmed away from the building. Sanitation is a big part of pest control in a hotel—making sure that the housekeeping staff is doing some deep cleaning, getting behind furniture, picking up any stray food. It's amazing, in some hotels, how quickly ants will appear when there is any food to scavenge. Ants are a big problem in hotels because of the food spillage and the different ways they can enter the building. Mice are also a challenge because they find nice warm areas, lots of runways, and a lot of food at hotels.

TREATMENT

Pest control technicians can apply exterior preventive treatments to keep out spiders, crickets, boxelder bugs, and ants. Often doing some exterior baiting for mice is needed and holds down their populations around the hotel. Mice really take to hotels and lodging, based on the number of calls we get. So we need to have some monitors and devices ready for them when they come.

Preventive treatments for bed bugs should also be done, in my opinion. A lot of companies just wait to get an infestation before taking any action, on the theory that prevention is impossible with so many people coming through. We have been doing preventive bed-bug treatments for years and find we can effectively make hotel rooms a hostile environment for bed bugs and prevent infestations. Maybe not all of them—but we can prevent a good number of them. We have cut infestations by 90 percent or more in some hotels, based on the trend reports we compile for our customers. We are applying a desiccant dust in the areas bed bugs are likely to travel and settle. It's applied very lightly so the guests don't see it, but it does a good job. If a guest brings in two or three bed bugs, there's a good chance they run into the dust and die. When we can see that a hotel that was averaging fifteen infestations a year is down to one a year or less after preventive treatments, the lodging industry should start to look at this seriously!

Other treatments and strategies will depend on the individual hotel's needs. Some will struggle with flies and others with ants. Programs should be built to address those specific needs or should be amended and addressed as they arise. However, good sanitation and exclusion practices often keeps those to a minimum, and detailed inspections and monitors should catch things quickly. A good pest-sighting log that is used by the entire hotel staff can also be invaluable to the pest control program, as staff often sees pest activity early in an infestation.

MONITOR AND EVALUATE

Placing monitors is important, but it has to be done with discretion in a hotel. To be useful, monitors must be where there's going to be pest activity, but we also want to hide them out of the way as much as possible. Most hotels don't want to see fly and mousetraps in their lobbies or guest rooms, so we look for creative places to put them, especially in breakfast areas or restaurants. Even bed-bug monitors can be placed with discretion if necessary.

Since bed bugs are one of the largest issues causing problems with hotel reputations, it is crucial to use a competent company that does a lot of bed-bug work and offers a decent warranty of three to six months. I see too many hotels getting either no warranties for bed-bug work, or very short ones—and they're paying to treat recurring problems in the same rooms! The chances of that same room having another guest bring in new bed bugs is pretty low, so if it's happening over and over, that's generally a sign that the population of bed bugs did not get eradicated completely. The company should be doing multiple treatments (if using chemicals) and should be verifying that bed bugs were eradicated to the best of its ability.

DOCUMENTATION AND EDUCATION

Hotel managers should be getting regular service records documenting which rooms were inspected, which rooms were treated, and what was done in those rooms. The pest control company should be recording what was found in all of the monitors and running trend reports throughout the year. That's how we tell if a particular area needs more devices or more treatment.

I think it's extremely important to educate hotel staff about the pest control program and especially about bed bugs. They're going to be the front line of defense. Housekeepers can do a good job finding bed bugs if they're trained well. All employees should be trained in general sanitation, what things they can help with, and what things cause pest problems. They should know what's being done by the pest management company and know how to report pest activity.

Front-desk staff should be trained in how to gracefully handle a pest complaint. They shouldn't simply brush it off and say, "No, we've never had any problems here. Of course we'd never have bed bugs." Instead they should be asking, "Why do you think you have bed bugs? What's going on?" That provides management with information and can also help calm the situation down. Staff can also say things like, "We take this very seriously. We want you to know that we have a detailed preventive bed-bug program and have done everything we can to detect and eliminate problems—but we will call a professional immediately to inspect the room and figure out what is going on."

The larger hotels should understand that they are often more vulnerable to pest problems because of their complexity. A pest control program shouldn't look the same for an eighty-room hotel as it does for one with six hundred rooms. There's a lot more going on: more laundry services, more employees, more turnover, and a lot more food (and garbage). They need to know that having someone show up once a month for an hour is not going to cut it.

SCHOOLS AND COLLEGES

Schools are a unique and complex environment for pest control because they are full of children all day and also store and serve large amounts of food. They are generally budget conscious and have little money for pest control, which can be a challenge when issues arise. We have to be creative to prevent and treat problems while avoiding using pesticides as much as possible. Even at the college level, we don't want to be applying pesticides while classes are in session.

Schools have lots of doors and windows, vents and pipes, and building maintenance can be lacking. There are lots of people, both children and adults, coming in and out. Food and garbage service can create sanitation issues. But having pest issues like mice and cockroaches is unacceptable because schools are responsible for the health and safety of the students. Given all these circumstances, schools must have an IPM program designed by a professional to be budget friendly, effective, and safe. Schools should take pest control issues seriously and find the budget necessary to have a quality program. They don't have to break the bank, but if they don't want to have mice running rampant, pest control should not be where the school tries to save money.

INSPECTION

Schools should have regular inspections of PVAs where there is food, easy entry, or sanitation issues. This means those areas need to be identified. The nonpest vulnerable areas can be inspected less often but should still receive proper attention. Building exteriors need regular inspection for any pest activity, such as wasps and other stinging insects. Some students will be allergic to those, so they have to be dealt with. We should have clear and detailed inspection reports that are given to the school of what was inspected and what was found.

We should rely, again, on as many nonchemical controls as possible: exclusion, keeping the pests out; sanitation, making sure things are cleaned up. Trash should be taken out frequently, the trash cans cleaned, and the garbage and dumpster areas cleaned very well. There should be no hiding places or habitats for pests: no tall grass, nothing stacked around the outside of the building, and no food debris in the cafeteria. Sometimes we have to move mulch or vegetation away from the sides of the building.

Vacuuming is a great way to do some nonchemical control. Heat treatments when necessary can be useful. But when chemicals are necessary, students shouldn't be present at the time of treatments. And then we should try to keep it to crack and crevice and spot treatments—not broadcasting the whole room. We're finding where the source is, treating the source, and treating some of the runways and entrances with spot treatments.

MONITOR AND EVALUATE

Rodent monitors should be placed in all the PVAs, making sure that we're seeing any issues, identifying problem areas, and catching any mice and rats quickly. Insect monitors in PVAs allow us to identify problems and eliminate them quickly.

Mice are the biggest pest problem in schools, but spiders, German cockroaches, ants, wasps, and other stinging insects are also huge issues.

School administrators should be getting inspection reports, monitor reports, and service reports whenever work is done. Trend reports of what's happening will help the pest control company stop pest infestations before they get out of control. The reports also help school officials track that the company is abiding by laws, regulations, and the rules on product labels.

EDUCATION

Pest control professionals should educate the school staff on what they can and should be doing. Sometimes the big issue is that teachers and other staff members don't know where or how to report a pest problem. They should know what pest control program is in place, what it looks like, and how it is protecting the students and them to create a healthy environment.

OFFICE BUILDINGS

Anywhere we have people working and eating, we can have some sanitation issues, and that often causes pest control problems in office buildings. Attractive landscaping also sometimes contributes. When plants are put up against the building and watered regularly, it creates moisture problems right outside, which can attract different pests like centipedes, millipedes, earwigs, thrips, and other insects. Inside plants can create issues with small flies and other insects. Plus, just the sheer number of people coming into office buildings can create problems as they unknowingly bring pests in.

INSPECTIONS AND TREATMENTS

Diverse office buildings are going to have different needs, so programs really should be individually designed. Some basic level of pest control should be performed to protect the health and comfort of the employees working there, but some buildings are going to need a lot more. The way to find out is by inspecting PVAs, the exterior, and pest-conducive environments such as open doorways. And it is becoming more common around the country to inspect office buildings for bed bugs. The more people working in a building, the more chances there are that one of them will bring in bed bugs, which can thrive not only in beds but anywhere people are sitting for long periods of time.

We should also be doing exterior preventive treatments for insects. Exterior rodent bait stations are often useful in office complexes. Multiple-catch mousetraps can be placed near any possible entrances. Making sure the building is well sealed, dealing with sanitation issues, and doing spot treatments around doorways and other entryways can also help with insects.

MONITOR AND EVALUATE

Rodent and insect monitors should be placed in PVAs, and bed bug monitors anywhere they are an issue. Along with inspection and service reports, these monitors let us see if there are any trends occurring.

It is helpful if management educates the employees in meetings or new-employee orientation sessions to keep their workspaces clean and free from food and debris. If the office is struggling with bed bugs, it should be discussed so employees know protocols, best practices, and how they can prevent problems. Human resources managers should have a plan so they are not scrambling to figure out what they are going to do if an employee either causes or reports a bed-bug problem.

I recommend having a written policy that tells employees, "If you have bed bugs at your home, please report it. If we know about it, we can put some prevention in place to help deal with it." The policy should be based on what your legal counsel says you can and cannot do or say—so make sure you check with him or her. But I've seen the best results with an open-door policy where we help employees get rid of the problem. We've had companies that pay for treatment at the employee's home so they don't miss work out of concern for bringing in bed bugs.

RETAIL LOCATIONS

Stores big and small are constantly getting deliveries that are often packed into storage rooms. These are often difficult to inspect well. Pest-monitoring devices may be appropriate in some retail spaces, but others may not want customers to see any pest devices. When some high-profile clothing stores in New York City had bed bugs several years ago, the media couldn't stop talking about it. But whether the problem is ants, mice, cockroaches, or any other pest, stores need a good pest management program. Not only do you not want customers to see pest activity, but many pests can also damage products and cost you money.

Any good program starts with a thorough inspection of PVAs, which may include storage areas that get little foot traffic. In addition to the usual concerns about good sanitation and sealing entrances, stores should look at storing goods off the ground and away from walls. Preventive treatments and rodent bait stations can be helpful. Some good monitors for retail locations are exterior bait stations, multiple-catch traps for mice, and insect monitors in PVAs. Ants and mice are probably the biggest pest problems for retailers, but management will know what they are dealing with if they get regular documentation from the pest control company on what's being found in the monitors.

While educating workers about proper sanitation and storage practices, retailers should make sure the employees know how to report pest activity. A lot of them don't.

That's a breakdown by industry of the challenges in pest control. The next two chapters take a closer look at each of the key pests to give you a better understanding of what it takes to control them.

CHAPTER 3
COMMON INTRUDERS

A Utah university (which shall remain anonymous) had a cockroach infestation for three years that several different companies couldn't solve. This was distressing because colleges are as proud of their campuses as they are of their football teams, which may be called the Lions, Cougars, Thunderbirds, or even Devils— –but never Roaches.

Thorn Pest Solutions had done some other work for the university, but we were meeting with new administrators, who were sizing us up. So I said, "Let's go take a look right now." We adjourned the meeting and did a full inspection. We immediately found German cockroaches in a janitorial closet; this was no surprise because it was wet and had drains as entryways. They told us, "This is where we've treated in the past . . . we keep treating this." So we started looking up and down and all around and found that the building architects had stacked janitorial closets on each floor, including one that had never been treated, and right behind these closets was an elevator shaft that had never been looked at and treated. We found garbage and debris there that was the roaches' food source, as was a nearby break room.

The main problem was they hadn't been digging deeply enough and looking three-dimensionally all around this building. Within a week and a half of our beginning treatments, the cockroaches were gone, and we had a program to clean up debris and monitor for any new activity. Looking 360 degrees for pests' food, water, shelter, and warmth is the way to get results. To do this, professionals must understand the biology of pests and then apply that knowledge to help eliminate the pests.

This chapter is going to cover various types of pests—not every single type, but the key pests that we deal with on a regular basis in urban environments.

GERMAN COCKROACHES

German cockroaches, the species *Blattella germanica*, is the most common cockroach in the United States, and they regularly infest structures where they can find food, warmth, and shelter. This is a domestic insect, not often found outdoors, and is typically brought in on cardboard boxes, grocery bags, food packaging, and used appliances such as microwaves. They are one of the most important pests we deal with because they spread filth and germs onto food, utensils, dishes, food-prep surfaces, and any other surfaces they happen to travel on. They can trigger asthma attacks and cause allergic reactions for many people.[1]

BRIEF EXPLANATION OF BIOLOGY AND LIFECYCLE

German cockroaches are light brown in color as adults. The juveniles are smaller, darker, and don't have wings. Adults are about one-half inch long. They have two dark stripes on the pronotum of their thorax, basically down the back. Males have a thinner body.

1 PhD Gary W. Bennett, *Truman's Scientific Guide to Pest Management Operations* (Questex Media Group LLC & Purd, 2010), 149.

Adult females carry their egg capsules with them until they're ready to hatch, which is nearly unique to these cockroaches. After the female carries that little egg capsule for twenty to thirty days, she drops it a day or two before the thirty to forty-five eggs inside hatch. Adults can live up to a year. German cockroaches are most active at night. If you see them during the day, it is often a sign that the population is very large or stressed—or both. They are thigmotactic, which means they prefer to be in tight, confined spaces like cracks and crevices during the day, and they tend to hide in places close to moisture, warmth, and food.

INSPECTIONS

Finding the harborages is important in knowing where to treat or place baits because the closer we can get the bait to the cockroaches the better, especially gravid (pregnant) females, as they don't generally venture far from their harborages. Inspections are critical to finding both harborages and hiding places, cracks, and crevices near moisture and food. Technicians should have a flashlight and knee pads for crawling in tight spots such as underneath sinks. Cockroaches spend up to 75 percent of the time hiding behind drawers, under countertops, behind stoves and refrigerators, in fuse boxes, and in appliances such as toaster ovens and microwaves. They can be behind the dishwasher, in wall voids, in cabinets, and in other places too numerous to list. Wherever there's a food source, you may find German cockroaches. Inspections should occur before every treatment. In multifamily housing, cockroaches often spread from unit to unit through plumbing, so it is important to also inspect surrounding units.

TREATMENT OPTIONS
BAITING

A lot of baits work very well for German cockroaches if used properly. Pest control professionals know to rotate baits with different active ingredients, use enough bait in enough places to control the population, and place it in the cracks and crevices where their harborages are.

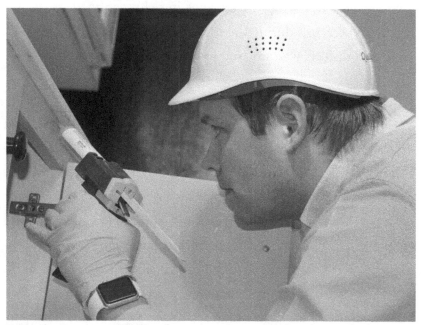

Baiting for German cockroaches.

You're not going to control German cockroaches if they have to travel far to get to the bait or if chemicals are interfering with their acceptance of baits. Tenants or restaurant workers don't need to remove dishes, utensils, and food before baiting as they would have to do in preparation for

Female German cockroach eating bait.

a chemical treatment. But with baits, multiple treatments are generally required with an inspection and monitor after every treatment to make sure we're getting the results we need. To see the results, the number of cockroaches caught in monitors should be counted and documented.

CHEMICAL TREATMENTS

Multiple treatments and follow-up are also generally required when using chemicals. Treatment should be directed into cracks and crevices, not just broadcast on baseboards and cabinets. You get much better results if your chemicals make direct contact with cockroaches rather than relying on them to come across residual pesticide later.

Treating cockroaches using ultra-low-volume machine.

When treating German cockroaches with chemicals, you want to hit the problem hard eliminating 95 percent of the cockroaches on the first treatment or cleanout. You sometimes hear companies refer to their cockroach treatments as cleanouts. Cleanouts often involve a treatment using nonresidual pesticide, delivered by a specialized ultra-low-volume machine that basically makes an aerosol fog that goes deep into cracks and crevices. Dusts constitute another good way of reaching into harborages; integrating some dust into the treatment program will help improve it significantly. Vacuuming up cockroaches is also a great supplement to chemical treatments to eliminate cockroaches quickly.

Any dishes, flatware, and cooking utensils must be removed or covered before chemical treatments. Don't just rely on people to clean them afterward. We know this is an issue—that tenants aren't going to do what we ask a lot of the time, especially when it involves completely emptying kitchen cupboards. That's a major disadvantage of chemical treatments compared with baiting.

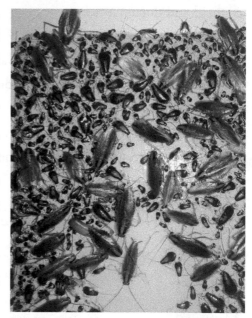

German cockroaches on a monitor.

Follow-up after chemical treatments should be scheduled to ensure complete eradication. Complete eradication should be your goal rather than wasting money with a company that wants to come in perpetually spraying baseboards month after month. After treatment is declared done, monitoring and a follow-up inspection a couple of months later will make sure a small population wasn't missed. Sticky traps make great monitors for German cockroaches. Professionals place them after the treatment and mark them with the date so you can see if any new cockroaches are caught. It is not uncommon to see a 95 to 98 percent reduction in the cockroaches and think they are gone, but five or six months later, they're back. You need 100 percent control. That's why you want follow-up inspections and treatment that comes with a warranty of three to six months.

COLD TREATMENTS

Cold treatments are starting to become a little more popular in the pest control industry and can be an effective tool for some pests if used correctly. The Cryonite machine shoots dry ice to eliminate pests. This can be lethal to pests that get hit directly. It has some advantages, including the ability to treat electronics, and it contains

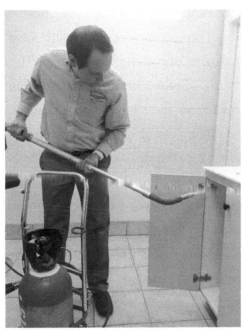

A professional uses the Cryonite treatment in a kitchen.

no pesticides, so it can be used in a variety of settings like restaurants and food-processing plants without worry of contamination. Also, pests cannot become resistant to it. Even with these advantages, our company has struggled to get great results using the Cryonite machine. If using Cryonite to eliminate German cockroaches, I recommend the company vacuum simultaneously, as some of the cockroaches will unfreeze and walk away. It should also be combined with baiting or spot treatments to be most effective. As with the other treatments, Cryonite applications require good follow-ups and monitoring.

ANTS

Ants are one of the most successful and interesting insects in the world. They can also be one of the hardest to control. Many ants are a threat because they sting or bite. They infest and contaminate food, can damage wood, and sometimes spread diseases. In other words, they're a big nuisance.

BRIEF EXPLANATION OF BIOLOGY AND LIFECYCLE

There are over eleven thousand known species of ants, most of which do not infest structures. They go through complete metamorphosis, which means egg, larvae, pupa, and adult stages. They have three distinct body segments. There's a head, an alitrunk, and a gaster. Where the gaster attaches to the alitrunk, that's called a pedicel. The pedicel may have one or two nodes, which is helpful for identifying the species. Ants have elbowed antennae.

Ants are social insects, and they live together in colonies with many individuals. In the colonies they have workers, which are all female. They have reproductive females, or the queens, and they have reproductive males. The reproductive males don't have a function other than to inseminate the queens. When all workers are the same size, the ant species is called monomorphic. When workers are different sizes, the species is called polymorphic. And when there are two worker sizes, it is called dimorphic. That also helps us distinguish what type of ants we're dealing with. Some ants nest in shallow nests in the ground, some in extensive ground nests, some in wood or hollow void, and some indoors. Proper identification is important to know where they nest.

INSPECTION AND IDENTIFICATION

Treatment should always start with a thorough inspection because proper identification is critical to determining the nesting sites and treatment options. The best way to eliminate ants is to find their colony, treat them directly, and find their food source. Different ants like different types of foods.

Inspections should be made outdoors where ants have been seen as well as indoors. Some companies do a good job eliminating nests indoors but neglect the outside, where there are more nests or satellite colonies. Odorous house ants and carpenter ants are two species that have those satellite colonies.

TREATMENT OPTIONS

Locating and treating the colonies directly generally leads to the best results. To do this, technicians have to dig around and lift rocks, boards, pots, or anything on the ground that can have a colony of

Baiting for ants.

ants underneath it. Prebaiting can be helpful. We put some kind of bait or jelly out and then do interior inspections or something else on the property. When we return to where we placed our bait, we look for ants in a trail that we can follow back to a colony. Nonrepellant pesticides are generally recommended for ant treatment. Spraying pyrethroids around your building and spraying the baseboards

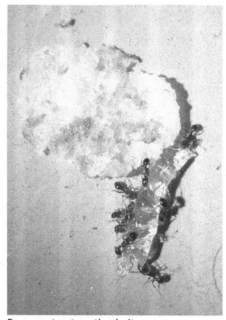

Pavement ants eating bait.

is often not going to control a colony, because often only a small percentage of the ants are leaving the colony to scavenge for food. Many types of bait are effective, but professionals will know how much to use and how to place the bait where the worker ants can find it. Worker ants do not feed while scavenging but take the food back to the colony where they share the food throughout the colony in a process called trophallaxis. Trophallaxis is the exchange of regurgitated liquids between ants. There needs to be enough bait taken back to kill off the whole colony. So follow-ups are generally recommended to determine whether we got complete control.

It's also important to control the aphids and other insects that produce honeydew around a property. These insects and their honeydew can be a major food source for many different types of ants.

RODENTS

Rats and mice are called commensal rodents, which means that they try to live among humans or in our buildings. They depend on humans for the shelter and food we provide them. They negatively impact our food, health, and comfort. Rodents are very good at spreading diseases and are responsible for killing more than ten million people in the last century alone.[2] Some of the diseases associated with commensal rodents are Weil's disease, plague, hantavirus, Lyme disease, and food poisoning, to name just a few. We're going to cover the two most common commensal rodents: house mice and Norway rats.

BRIEF EXPLANATION OF BIOLOGY AND LIFECYCLE

Mice have an amazing ability to adapt to different conditions and can be found throughout the world. Their small bodies allow them to enter tiny openings and occupy confined spaces. If food and shelter are available, they are very prolific. They need very little water to survive. They feed on a wide variety of food and have small territories, going no further than about a ten-to thirty-foot radius. They're nibblers. At night, when they are most active, they'll make constant little trips to get just a few micrograms of food at a time. Mice are excellent climbers and can scale walls easily. A single house mouse produces fifty to seventy-five droppings and hundreds of microdroplets of urine a day. Their behavior can be erratic and unpredictable, which sometimes makes them difficult

---

2 Robert M. Corrigan, *Rodent Control: A Practical Guide for Pest Management Professionals* (Cleveland: GIE Media, 2001), 15.

to control. Some mice show little interest in our traps, and some don't even travel along walls.

The Norway rat was introduced to the United States in the 1770s and has since spread to many areas. Norway rats are ground-dwelling mammals that dig burrows. Rats regularly infest structures where they can find warmth, food, and water. Unlike the house mouse, they need a steady supply of water. Rats will scavenge for a variety of food, including pet food, birdseed, dog feces, vegetable gardens, and a wide variety of human-generated refuse. Norway rats will travel about twenty-five to one hundred feet from their burrows, or less, if food and water are available.[3]

INSPECTION AND IDENTIFICATION

Rodent control should always start with a thorough inspection to determine the extent of the infestation, where they are living, and where they are going. To make an effective treatment plan, we have to find the high-traffic areas, entry points, and food sources. Mice need only a pencil-thin hole to get into a building. The inspection is looking for rodent droppings, runways, grease marks called sebum, and urine stains. Droppings become hard within a couple of hours, making it possible, wearing gloves, to pick them up, squish them, and tell if they're old or fresh. That's good information to have. A trained professional can use black light to find urine spots to identify entry points and high-traffic areas. Understanding rodent behavior allows us to focus on areas that provide warmth and food and their preferred corners, shadows, and caves.

---

3 Robert M. Corrigan, *Rodent Control: A Practical Guide for Pest Management Professionals* (Cleveland: GIE Media, 2001), 47.

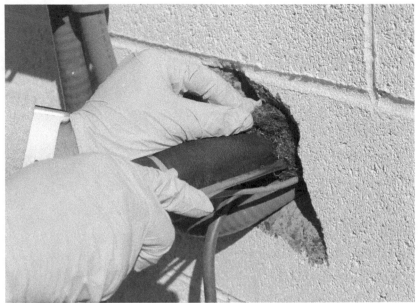

A professional seals up a building to exclude mice and other pests.

TREATMENT OPTIONS

Sanitation—removing as many food sources as possible—and sealing up the building as best as we can are the two most important steps toward controlling rodent populations. Removing clutter inside and right outside a building removes hiding places and harborages. Stacking pallets and products right up against the walls in commercial facilities is a mistake because there should be an inspection aisle at least eighteen inches wide.

Trapping is a great treatment option. There are snap traps for a single catch. There are multiple-catch traps, and there are glue boards or sticky traps that can be used to control these populations, especially

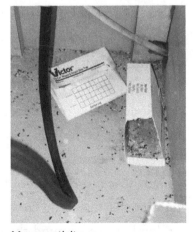

Mouse activity.

73

mice. In our experi-
ence, snap traps, with
proper placement, have
been more effective
than the glue boards
or the sticky traps. But
we use a wide variety of
traps and bait to entice
leery mice. When
using sticky traps, we
place them flat on the
ground—open versus
folding them up. They
catch more mice flat.
Often when we come
into a situation where
there has been a failure

Sticky trap for rodents.

to control mice, we find that not enough devices were placed. If
you have an apartment kitchen infestation, you want to see fifteen
or twenty traps, not three. The other common reason we see failure
to control rodents is placing traps along the wall and not putting
the traps where the mice live and travel. You must take the traps to
the mice and not just set traps along the wall and hope mice come.
Some mice are just not curious.

Rodenticides placed in tamper-resistant bait stations outside
buildings can be very effective in reducing rodent populations, but
proper placement is crucial. The bait generally contains an antico-
agulant that causes internal bleeding but kills rodents humanely.
Bait stations must be in areas of high activity and runways, not just
randomly placed around the building. We look for natural places

Exterior rodent bait station.

to hide from predators or for signs of rodent activity. Sometimes baiting fails because the pest control professionals allow the bait to be contaminated by pesticides they work with or other products. Rodents will stay away from chemical odors, including cleaning supplies or even nicotine smells that get on a bait station placed by a smoker who doesn't wear gloves to handle it. Bait that is old and rancid, moldy, or infested with insects will not entice rodents as well as fresh bait will. So the bait should be changed every four to six weeks at a minimum. It may have to be refreshed once or twice a week to control a large population of rodents, or it will be all gone before some get to eat it.

FLIES

Many flies are synanthropic, meaning that they live in close association with humans. Many people simply see flies as annoying, but they are not just annoying—they can also be detrimental to our health. Because of their filthy habits, flies are flying infections and spread over one hundred diseases, including typhoid, paratyphoid, cholera, dysentery, pinworm, hookworm, and tapeworm. They pick up disease-causing organisms from garbage, sewage, feces, dead animals, or other filth and deposit them on our food. Because of this, control in food and health-care facilities is very important.

BRIEF EXPLANATION OF BIOLOGY AND LIFECYCLE

Flies (scientific name *Diptera*, Greek for two wings) go through a complete metamorphosis with an egg, a larvae, a pupa, and an adult stage. Large flies, like houseflies, blowflies, and flesh flies, lay their eggs in warm, moist materials such as manure, human or dog excrement, garbage, dead animals, and decaying vegetable matter. Large flies most often breed outside, while small flies, such as fruit flies or phorid flies, typically breed indoors. Flies cannot eat solid foods, so they liquefy their food with regurgitated saliva and then suck up the liquid. So if a fly is feeding on your food, you are also getting this regurgitated saliva composed of liquid the fly may have ingested in its not-so-pleasant breeding grounds.

INSPECTION AND IDENTIFICATION

Inspections are focused on locating and eliminating the breeding sources, a vital step in eliminating the flies themselves. Food scraps, spills, and dead animals are examples of the many sanitation issues that can attract flies. In addition to cleaning those up, we are looking for fly entry points, such as missing screens and poorly sealed doors. We should be inspecting the dumpsters, which should be at least fifty feet away from the buildings and regularly washed so gunk and debris don't build up on the bottom. Lids should be closed and tight fitting, which often they are not.

Different types of flies have different breeding sources. So identifying the type of fly can point us in the right direction for control methods.

TREATMENT OPTIONS

A program with multiple treatment options usually gets the best results. The first step with large flies is simply keeping them out with window screens, doors that fit tight, etc. Elimination of breeding sources found in the inspection is imperative. A lot of fly baits work very well. Users have to read the label and pay attention to where the baits can be safely deployed. But placing fly bait stations in critical areas, such as around the outside of the building or around dumpsters, can be effective in reducing fly populations outside.

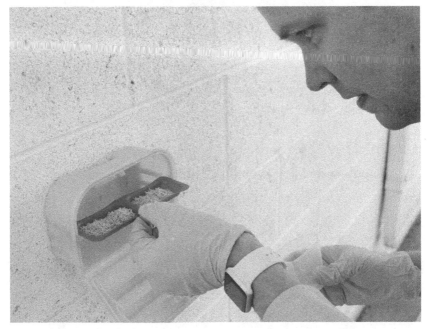

Servicing an exterior fly bait station.

Insect light traps are a great component of good fly management, if placed properly. These are traps that use lights to attract flies onto glue boards. (A full explanation and guide is in the book's appendix.) Typically, these should be inside a building, two to five feet high on a wall, not above food preparation and not in an area where they would attract flies from outside. Some restaurant or food-plant managers who have a light trap are confused about why they still have flies. The answer is that they may need multiple fly lights in combination with other strategies described here.

A professional inspects an insect light trap.

Glue sticks and flypaper are another control strategy, at least for nonpublic areas. It would be fairly unsettling for restaurant customers to see fly sticker papers full of dead flies. A fairly new product from manufacturer FMC called EndZone is a clear sticker with a sugar bait matrix that attracts and kills flies shortly after they land on it, without being a sticky paper. It can be used in a lot of places, including on trash cans and windows, as a great supplement to a good fly-control program.

Residual pesticides can be used in places where flies are likely to land. There are some nonresidual pesticides that you can combine with an attractant and make a fly-kill zone. ULV treatments are basically a cold fog of nonresidual pesticide that floats up into the air and can knock down adult flies if that population gets bad. Professionals need to obey the label and make sure they are not contaminating food, utensils, or dishes. These are just the basics of fly-treatment

Glowing fly stick.

options. Your pest control professional can consider all of these methods and more, choosing the correct control strategies after identifying the types of flies involved.

SUBTERRANEAN TERMITES

Termites need the cellulose in wood as a food source. That's what makes them such a destructive pest in structures containing wood. The NPMA estimated that termites did $5 billion worth of damage in 2005, with an additional $2 billion paid for remediation.

BRIEF EXPLANATION OF BIOLOGY AND LIFECYCLE

Termites go through gradual metamorphosis. They are social insects, which means there are different roles and responsibilities within the colony. Typically, termites have reproductive and soldier castes, and many termite colonies have a worker caste. The workers are the ones that can do quick and devastating damage to wood structures while the soldiers protect the colony.

INSPECTION AND IDENTIFICATION

Termites are sometimes confused with ants, but there are clear identifying characteristics. Termites are thick waisted, where ants have thin waists. Ants have a distinct elbow in their antennae, and termites do not. The alate, or reproductive caste of termites, has two pairs of wings equal in length. Reproductive ants have two different sizes of wings. The workers among both termites and ants have no wings.

Sometimes termite activity is completely obvious. You see mud tunnels going up the foundation of the wall, and you can easily see that there's been some termite activity. Sometimes people see termites inside their homes, but other times it can be really difficult to find out if termites are invading a structure. The determination can require a lot of effort and knowledge. Professionals have to move slowly, using a bright flashlight and looking very closely at every potential problem area in a structure. So not only do they need to know about termites but also about construction to understand how the termites are getting in at the footings and floor joists. Wood that is damaged is hollowed out along the grain and often has bits of soil in the gallery. Poor inspection is the main reason for treatment failures.

Someone doing inspections may carry knee pads, a good flashlight, knife, screwdriver, and small hammer. Tapping on wood with the handle end of the screwdriver or small hammer can help find hollow spots. Sometimes you can even disturb the soldier termites enough to hear them start to move around.

Subterranean mud tube.

Mud tubes are probably the most recognizable sign of subterranean termites. Living underground, needing moisture, and not liking exposure to the air, they build tunnels to travel to other places to feed on wood. They can build several inches of these tubes each day, leaving brown mud-like lines or galleries on the surface as a surefire sign of termites. If the tubes are dry and brittle, there may not be active termites, which generally leave moist and supple tubes.

TREATMENT OPTIONS

There are too many different types of buildings and treatment scenarios to cover them all in depth in this book. So I'm just going to briefly familiarize you with the most common possibilities. It has become increasingly common to place bait stations in the ground around the structure. These have some type of attractant—wood or cellulose material—that termites naturally scavenge for. The bait stations have to be checked regularly for any activity, in which case you place a bait insecticide inside something termites like to eat that they can take back to the colony to kill it off.

This type of treatment doesn't require digging trenches or drilling into the structure, as is necessary with chemical treatments. We can do it all from the outside of the building, and it's more environmentally friendly than putting fifty gallons of termiticide down around a home. It takes a lot less labor and is less toxic than liquid pesticides. Generally, these baits have very low toxicity and are safer to be around. The disadvantage of baits is that they take longer to eliminate an entire colony. And baits are not easily placed underneath slabs of buildings and in wall voids where termites are often located. So sometimes there's a challenge in getting the baits to the termites.

There are lots of different types of chemical treatments. These may involve trenching around the outside of the building, drilling into slabs and porches, or a technique called *rodding*. That involves a high-pressure injection through a rod that we stick several feet into the ground in order to get the product deep to reach the bottom of the foundation. Indoor treatments include drilling into slabs, foaming around bath traps and utility lines, and treating wood directly inside of a home.

Chemical treatments are hard work, and they need to be done right to create a continuous barrier of insecticide around the foundation and underneath the building. Professionals doing termite work are going to have to be very knowledgeable not only about termites but also about building construction and the safe use of chemicals. It's hard work, but chemical treatments can be very successful. There are some great products on the market—especially some of the nonrepellent pesticides that can keep termites out for a long time. Just make sure you find somebody who's competent in termite work.

STINGING INSECTS

According to the American College of Allergy, Asthma, and Immunology, over two million Americans are allergic to stinging insects such as bees and wasps, and over five hundred thousand people visit hospital emergency rooms every year in the United States because of stings. Stinging insects have venom that attacks the central nervous system or destroys cells and tissue. People suffer not only from the effects of the toxin but also from allergic reactions that can range from mild to very severe and life threatening. There are approximately one hundred deaths in the United States annually from bee and wasp stings.[4]

BRIEF EXPLANATION OF BIOLOGY AND LIFECYCLE

Bees and wasps both belong to the order of *Hymenoptera*, the same order as ants. They go through a complete metamorphosis with egg, larva, pupa, and adult stages. There are solitary wasps, which just go around doing their own thing and pose little threat, and social wasps—the hornets, paper wasps, and yellow jackets that live in colonies and are a major threat. That's the type we'll going to cover in this section.

INSPECTION AND IDENTIFICATION

Treatment starts with an inspection of the property and the surrounding areas. Pest control professionals should ask people at the building where they have seen activity and collect specimens if possible to identify the species and determine the best treatment strategies. It is important to try to locate where they live and if there are food sources around that are attracting them like dumpsters and trash or honeydew-producing insects.

4 Arnold Mallis, *Handbook of Pest Control*, ed. Stoy A. Hedges, B.C.E. (Valley View, OH: GIE Media, Inc., 2011), 826

TREATMENT OPTIONS

Technicians may need a bee suit, which includes coveralls, a hat with a tight-fitting veil, and high-sleeved canvas or leather gloves. They may be applying insecticides three different ways, with:

1. A duster, which might be electric or manual—squeezing a bulb to release dust

2. Aerosol cans of pesticides that provide really quick knockdown

3. A backpack or hand-canister sprayer with a residual pesticide

Other equipment includes exclusion material to seal the wasps and bees out of buildings, a spatula, a pole to remove nests, a ladder, and a cordless drill. We sometimes need to drill into the walls to get access to these nests. Species that nest in the eaves and behind wallboards can be treated by injecting that void with a vdust insecticide. It works very well to kill off the whole colony quickly.

Food in dumpsters attracts yellow jackets and wasps. Dumpsters should be kept clean, well sealed, and at least fifty feet away from the building. Aphids, scale, and other insects that excrete honeydew must be controlled with an appropriately labeled pesticide because they attract paper wasps, yellow jackets, honeybees, and ants. When people see wasps or yellow jackets swarming around a tree or bushes, they think that's where the nest is, but it's really just them trying to eat the honey-dew.

Paper nests of bald-faced hornets and aerial yellow jackets are best treated at night because the colony's foragers are in the nests at night. Treating with a dust is often the safest and most effective strategy, as it doesn't stir them up quite as much, and it's really effective at killing these insects. This can be done by putting dust into the entry and exit holes or by also puncturing the sides of the

Dolichovespula maculata wasp nest in a tree [L] and after removal [R].

nests to fill them with dust. Paper wasps can easily be eliminated with a fast-acting residual aerosol insecticide. It is important to remove the nest and treat the area with the residual pesticide to prevent wasps from building there in the future.

STORED-PRODUCT PESTS

"Stored-product pests" is a term encompassing a variety of insects that together damage more than 10 percent of the world's grain production. There are too many stored-product pests for us to talk about all of them here, but we're going to mention a few of the most common species along with some basic guidelines for inspection and treatment.

BRIEF EXPLANATION OF BIOLOGY AND LIFECYCLE

Internal feeders (species like the rice weevil, the granary weevil, the angoumois grain moth, and the lesser grain bore) lay their eggs inside grain, so the larvae hatch and feed there. If they feed on the outside of the grain, they are external feeders like the drugstore beetle, cigarette beetle, trogoderma, and the Indian meal moth.

And then there are scavengers that feed on grain only after the seed coat has been broken, like the confused flour beetle, red flour beetle, flat grain beetle, saw tooth grain beetle, and the Mediterranean flour moth.[5]

INSPECTION AND IDENTIFICATION

As noted in the last chapter's section on food-processing facilities, professionals should be looking inside and outside for conditions conducive to pest infestations. They should also look for insects or signs of insects for identification. Proper identification is needed to choose the best treatment methods because there are so many different types of stored-product pests. Monitoring tools, such as ILTs and pheromone traps, are very useful in detecting stored-product pests and determining the severity and extent of infestations.

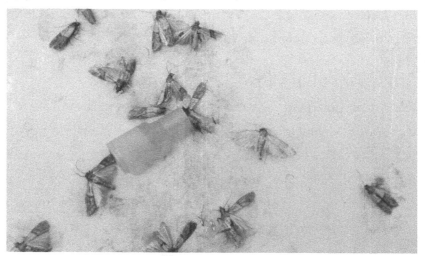

Indian meal moths caught on a pheromone trap.

TREATMENT OPTIONS

Good sanitation and storage practices are critical to eliminating and preventing stored-product pests. Food often gets trapped in

5 Gary W. Bennet, John M.Owens, and Robert M.Corrigan, *Truman's Scientific Guide to Pest Control Operations* (Advantar Communications, 1988), 271–290.

large machinery in food-processing plants and can be a great food source for these pests. So food facilities need a program to clean equipment and spills, keep doors closed, and seal holes and gaps. Food-handling locations should practice FIFO, or first in, first out, meaning rotating products so they are not sitting as long, which increases the chances of infestation. There should be inspection aisles of at least eighteen inches along the walls. Vegetation should be kept away from the building. All incoming trucks and materials should be inspected for evidence of pests.

ILTs can be a way to catch and eliminate some of the stored-product pests but also can be very helpful in identifying which of them are present. All insects caught on ILTs should be documented with the species and the number caught, and this should be analyzed for trends.

We usually try to attack every life stage of these insects. This involves locating all infested foods to eliminate the larva and eggs. Crack and crevice treatments can be done to try to eliminate the larvae that are leaving the food to pupate. And then traps are placed to eliminate as many adults as possible.

Heat can be a successful way to control stored-product pests. This practice has become more common. The temperature of the facility is raised to 120 to 150 degrees Fahrenheit for approximately twenty-four hours, and this is to make sure that that heat penetrates deep into those cracks and crevices and kills all of the insects. If pesticides are needed, we should always obey the label and ensure we're not contaminating food. A low-pressure spray into cracks and crevices is a recommended treatment in food facilities.

STRUCTURE-INFESTING BEETLES

There are millions of species of beetles, which are quite diverse and make up the largest order of insects. They are a big problem in commercial pest control because they feed on and infest a wide variety of food, and some can destroy wood. Because of the diversity, misidentification happens frequently, and there are a lot of different control techniques.

BRIEF EXPLANATION OF BIOLOGY AND LIFECYCLE

Beetles live in almost every habitat imaginable. They belong to the order of *Coleoptera*, which is Latin for hard wing—a good description. They go through a complete metamorphosis, which means there's an egg, a larva, a pupa, and an adult stage. Beetles feed on a wide variety of plants and animal materials. Some prey on insects, some scavenge for food, and others feed on mold or fungi.

INSPECTION AND IDENTIFICATION

The easiest way to identify if an insect is a beetle is by looking at its back. A beetle's wings cover its entire back area, meeting in the middle of the back and forming a straight line down the back. Different beetles require different control strategies, so it's important to correctly identify the species. Sometimes this requires collecting samples and sending them to an entomologist or agricultural extension office. Inspection and identification should both help determine where the beetle is living and its food sources.

TREATMENT OPTIONS

There are too many different beetle species to specify here which treatment options are best suited for each, but you should be aware of some common methods. These include the same sanitation and exclusion measures covered above in the section on stored-product pests. Heat treatments are becoming more popular, as heating the area containing the beetles to 120 to 150 degrees will kill all life stages. Freezing and vacuuming up the insects are two other non-chemical-control methods. With insecticides, a low-pressure spray is used in cracks and crevices to avoid contaminating food sources. Filling wall voids with dust or foam also can treat areas where beetles are living. Spot treatment involves targeting one to two feet of an area that is an insect-activity hotspot. Wood-boring insects may be treated directly with insecticides. Larger areas, such as a grain silo, may require fumigation. ULV space treatments involve a cold fog that fills the area.

Follow-up is needed to determine if treatments were successful or if further action is needed. Some great monitors are available for this follow-up.

SPIDERS

Spiders play an important role in controlling insect populations, and we should be grateful for the work they do. But when spiders get into homes and businesses, they may need to be controlled. Sometimes they're just a nuisance, but some people are really afraid of them, and there are venomous species in parts of the country.

BRIEF EXPLANATION OF BIOLOGY AND LIFECYCLE

Spiders are predacious, meaning they are predators feeding on insects. They also feed on other spiders, and sometimes even feed on small mammals and reptiles. Spiders have eight legs, two body regions, which are the cephalothorax (a fusion of the head and the thorax) and the abdomen. They have two appendages called pedipalps by the mouth that help capture food and bring it into the mouth; if you see a spider and think it has ten legs, those front two appendages are actually its pedipalps. Almost all spiders spin webs. Many have different types of webs. Not all spiders use webs to catch their prey. Many leave their webs to hunt their prey. Many spiders have eight simple eyes. And eye patterns are important because they help tell us what type of spider it is.

INSPECTIONS AND IDENTIFICATION

Good spider control starts with a good inspection to learn what types of spiders are present and why. Are there food sources such as other insects around? Are there exterior lights and other landscaping items, such as heavy vegetation against the building, that are attracting these spiders? Lights that are left on all night are welcome signs for insects, which bring in the spiders.

There are tens of thousands of species of spiders. So, identifying individual species can be very difficult. But if we can at least identify approximately what group the spider belongs to and can determine whether the species is harmful, that can really help.

Different groups of spiders that we're looking for during inspections include:

- jumping spiders, for which the scientific family name is *Salticidae*
- crab spiders, the family *Thomisidae*
- ground-hunting spiders, the family *Gnaphosidae*
- wolf spiders, the family *Lycosidae*
- orb-weaving spiders, the family *Araneidae*

- tarantulas, the family *Theraphosidae*
- comb-footed spiders, *Theridiidae*, which includes the most commonly known of that family, the black widow
- recluse or violin spiders, the family *Sicariidae*, which has the infamous brown recluse[6]

Those aren't all the families of spiders, but they are the main ones that we deal with in the pest control industry in the United States. Each one requires a slight variation of best practices for controlling them. For example, wolf spiders are active hunters. Knowing that they're going out trying to hunt down their prey makes them a different kind of target than a recluse spider. So, identifying which family or group they belong in can really help in treatment techniques, how we're going to attack the problem, and the best ways to do it. Knowing roughly which species we're dealing with helps us know whether they are living mostly outdoors or indoors and what type of food they are going for.

TREATMENT OPTIONS

The first step is removal of the spider webs. That is most critical for black widows. They tend to have a hiding hole that they can go into when their web is sprayed, but if it is removed before the treatment, the black widow will come out to try to rebuild her home and will run into the product. Exterior treatments with residual pesticides should target places where spiders are likely to build webs. Eliminating the other insects that the spiders are eating will help to control the spiders. Exterior treatment of the foundation with a residual product helps, as does spot treatment indoors

6 Stoy A. Hedges, *Anthology: The Best of Stoy Hedges* (Cleveland: GIE Media, 2001), 246–249.

and around windows and doorways. If a building has a crawl space underneath it, that generally needs to be treated. Harborages near the building, such as log piles, long grasses, and anything right up against the building should be removed, and entry gaps in the building should be sealed.

For follow-up, some simple monitors can be placed on the grounds to see if treatments were successful or if there's heavy activity. They can also tell us what types of spiders are active. The results of the monitoring can inform us when we need to treat again and what else might need to be done.

CHAPTER 4

B2B SERVICE (AS IN BYE TO BED BUGS)

The elderly and disabled residents of a suburban Salt Lake City apartment complex had been through a lot, and so had their management company. They had bed bugs throughout the 140 units and hallways of the two midrise elevator buildings. Some of the residents had been put out of their homes for a full day for heat treatments, not once or twice, but three, five, or even seven times. When we took over the battle, the attitude of the residents was like, "Yeah, we've been through this before. We know the drill. It's not going to make any difference."

This was a renovated apartment complex with amenities including a chapel and physical-therapy facilities. Managers told us they had spent $90,000 with another company in the prior year treating numerous units in a futile effort to control the bed bugs. I looked them in the eye and said, "The first thing I want to do is inspect all of the units myself with my canine—Radar—to find out what's really going on." They said, "Is that necessary? We already know. The other company just went through last week and they found bed bugs in eleven more apartments." I said, "Yeah. I know it's spending a couple more days doing this, but I really think it's wise." So we went through all 140 units. We found fifty-two units that still had live bed bugs. Then we sat down with the managers and designed a comprehensive program, which included heat treatments in the units with live bed bugs and preventive treatments in all the other units. We addressed the hallways and then designed a program to follow up and do inspections quarterly to make sure things were under control at the property.

The results were fantastic. Within three months, they had less than a 1 percent infestation rate. We've been doing quarterly inspections on all 140 units for the past three years, maintaining the same under 1 percent infestation rate at this property, which has gone from spending $90,000 to less than $15,000 a year on this. The manager is happy and so are the tenants.

HISTORY

Bed bugs were mostly eradicated throughout the United States in the forties and fifties. One of the main reasons was the emerging use of organophosphate pesticides, including DDT, a pesticide that has since been scrutinized and even banned due to its effects on humans and the environment. Even in hospitals, they used to spray down the bed with DDT, flip the mattress over, spray it on the other side, and really coat everything with the chemical. That really did a number on bed bugs. People were vigilant too, constantly treating and watching for bed bugs. In the mid-1950s, however, bugs began developing resistance to DDT, and the National Pest Control Association started recommending other insecticides, such as malathion. They really licked the rest of the bed-bug problem, so there were very few cases in the United States until they reemerged in the 2000s.

There were several reasons for the reemergence, the largest being much more widespread and frequent travel. Other reasons for the reemergence include:

- a boom in the bed bugs' food supply—the human population
- more judicious use of pesticides, treating specific rather than broad areas
- bed bugs developing resistance to modern pesticides
- the buying of used furniture becoming more popular

Bed bugs have been around since at least the beginning of the written word, but they've now spread a lot more than ever. A big part of the problem was that people born after bed bugs mostly disappeared knew little beyond the old saying, "Nighty night, sleep tight, don't you let those bed bugs bite." People did not learn about bed bugs in school and couldn't identify them. Some had them in their homes for a long time before they recognized what was going on. All these things came together by the mid-2000s to allow bed bugs to take the United States and other countries by storm, starting in the major cities but now in small, rural towns, as well. It doesn't look like the problem is going away anytime soon; it actually continues to get worse year after year. We originally saw bed bugs in multifamily housing and hotels. Now we've started to see them in stores, restaurants, libraries, offices, city buses, police stations, fire stations, and clothing stores. We are not currently winning the war against bed bugs.

BED-BUG BIOLOGY

The purpose of this brief biology lesson is to make you a bit wiser about bed-bug behavior. The first thing we should talk about is that you are their food source. They feed on blood anywhere from three to twelve minutes one to three times per week. They prefer human blood, but they're opportunistic and will feed off a dog, a cat, a mouse, or a hamster. Females lay two to three eggs per day—they are egg-laying machines. Those hatch in seven to ten days. They have five nymph stages and are adults in one to two months, as long as they have a food source. Their average life span is six months to a year and a half. They can live a long time without feeding, some as long as six to twelve months. Bed bugs are cryptic and spend most of their time hiding in cracks and crevices waiting for their food—you—to return.

Bed-bug bites are very difficult to identify, and not everyone has the same reactions. Determining whether or not bed bugs are present should not be from bites alone. You need to find the bugs. Some people are not allergic to bed-bug bites, while others are very allergic and get red welts. People can also have a delayed reaction where the bites do not show up for up to fourteen days.

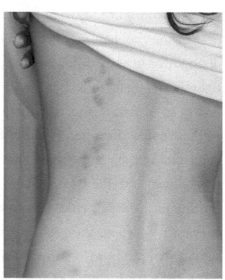

Bed-bug bites.

Bed bugs are synanthropic, meaning that they live in close association with humans. They do not spontaneously generate, and therefore, they must be brought into a home or property. They are most often brought in on used furniture, luggage or bags, or from hitchhiking bed bugs on a visitor or houseguest. When someone lives in a place that has bed bugs, especially heavy infestations, it is not uncommon for them to have bed bugs in their clothes, shoes, or belongings.

Bed bugs have a unique way of copulating. It is called forced insemination. Basically, the male has a harpoon-shaped genital that pierces the female cuticle in the abdomen. Women always think this sounds very pleasant. Females get copulated with often,

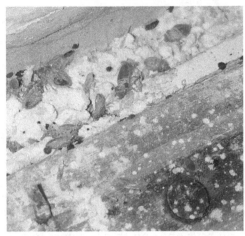

Shedded skin from bed bugs.

generally after a blood meal, as males prefer to mate with fat females, so they almost always have enough sperm to continuously produce fertilized eggs. A female has two ovaries with fourteen ovarioles—so she always has plenty of eggs to lay. One adult female can cause an infestation. Once she gets brought in, she starts laying eggs. Those eggs hatch, and in one to two months, they are adults and begin to

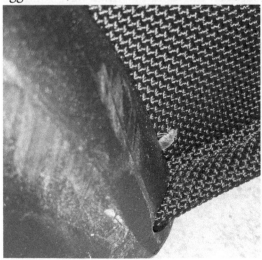

breed with each other, causing an infestation. These behaviors are part of why bed bugs are so successful.

Bed bugs on luggage.

PREVENTION

To prevent bed bugs from establishing themselves on your property, the three main things to do are:

1. Educate your staff and tenants on how to identify bed bugs and how to avoid infestations.
2. Schedule and maintain professional inspections.
3. Schedule and maintain preventative treatments.

One of the big problems we're facing involves tenants not reporting bed bugs. This may be because they fear that they're going to get evicted. They fear they're going to have to pay for the treatment. They're embarrassed. There are lots of reasons, but if you have one unit that becomes heavily infested, it can spread. The bed bugs can travel through the hallways or the walls themselves, depending on the construction, or they can be spread by tenants visiting neighbors. We should be teaching tenants how to prevent that.

In its metamorphosis from nymph to adult, a bed bug grows and darkens from a light tan to a rustier brown color. And there is a small shape difference between the females, which have rounded behinds, and the males, which have pointed rear ends. Otherwise, bed bugs are similar in appearance, so tenants can be taught to spot them.

Unfortunately, most people can't tell a bed bug from a baby cockroach. The media frequently covers bed bugs, and they occasionally show pictures, but most stories are more about the scope of the problem and the people struggling with bed bugs. Unfortunately, the coverage does not educate the public about the biology and where and how to look for them. I still get calls all the time from people who have no clue what bed bugs are, how to identify them, or how they could get them.

That bring us to the places where bed bugs are really entrenched—multifamily housing. Such facilities need to have frequent inspections, once a year at a bare minimum. The more frequently we go out and inspect these properties, the more likely we can catch infestations when they're small, easy to control, and haven't spread all over the place. That's the key to getting control of bed bugs.

Heavy population of bed bugs.

There are also preventive treatments for bed bugs. Preventive treatments won't stop every infestation, but they work really well. These preventive treatments combined with inspections can save an apartment complex a lot of money and spare the tenants a lot of hassle. Silica gel powder is applied in potential bed-bug harborages and places where they are likely to travel—along a baseboard under a bed, along the bed frame, on bed legs, behind the headboard, underneath the box spring, in the seams of couches and chairs.

This dust can literally stop an infestation when a few bed bugs are brought in. Or it can at least slow down the infestation. Silica gel, used for decades to control moisture, is very safe for people and pets but can kill bed bugs when applied properly. You're going to see a lot more of these treatments in the future because increased use is backed by research published in 2014 by Michael Potter, professor of entomology at the University of Kentucky.

Dr. Potter used paintbrushes and makeup brushes to apply the silica gel. A very small amount of dust was lethal to bed bugs even if they were on it for just a few seconds. Dusts must be applied the right way in the right places and not overdone. A bed bug is not going to cross a pile of dust. We use a specialized gun that applies the dust in a nice, light application. If the dust is simply thrown around the bed, it won't be nearly as effective.

DETECTION

We've discussed using constant vigilance and inspections to locate bed bugs early. Beyond the fact that many tenants don't report problems, there are several reasons why a pest control professional should be involved. Different properties require different protocols: visual detection,

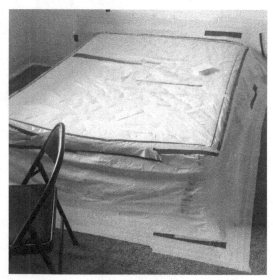

This client taped up their bed to stop the bed bugs from biting.

specially trained dogs, and monitors. I prefer using multiple methods, and I will tell you all about doing visual inspections alongside my dog in the next chapter. It's very low tech, but we have to know where to look. I get on my hands and knees and crawl under beds with a flashlight, carrying little tweezers and vials.

Radar checking for bed bugs

Bed bugs like rough materials, like unfinished woods, and can be found in screw holes in a bed frame or around the staples and creases under the box spring. We typically are more likely to find bed bugs on the headboard side of the bed, and underneath the box spring is the number-one place we find them. We also find them often on the backsides of wooden headboards, on wooden bed frames, and in whatever chair or couch has the best view of the TV, as it is often occupied by a warm-blooded human. These are the most common places to find bed bugs, but we also find them in random places throughout bedrooms and living spaces, which makes treatment more difficult. Good bed-bug inspections require thoroughness, good understanding of bed bugs, patience, and a keen eye.

There are several monitors that can give decent results in detecting bed bugs. The best are probably the pitfall traps like the ClimbUp Interceptor or the BlackOut Bed Bug Detector. They are simply disks that can be placed under bed and furniture legs, and when a bed bug falls in they cannot get out. They are very low tech but have pretty good results. We have also had decent results with the SenSci Volcano. It's about three inches square and looks like a small black volcano with an open top. A lure developed by Rutgers University scientists, SenSci Activ, is placed inside, and it has a clear bottom so

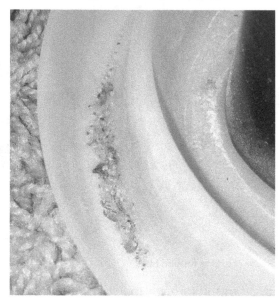

Bed Bugs and carpet beetle larvae caught in a ClimbUp Interceptor.

we can see the bed bugs trapped inside. The monitors need to be rebaited every ninety days but can be cost effective when we can't otherwise locate bed bugs in a unit where the tenant keeps complaining. A lot of the sticky-trap type monitors are not effective. Bed bugs can often pull themselves loose from sticky traps.

Another problem with amateur detection is misidentification. Bed bugs are part of a family of insects called cimicids. There are about ninety different types of cimicids, and some look similar to bed bugs. One of the main differences, for example,

Swallow bug.

between a bed bug and a bat bug is that bat bugs have slightly longer hairs near their head. Swallow bugs have a slightly hairier body, are a little bit smaller, and are grey rather than brown. We have also seen Mexican chicken bugs misidentified as bed bugs.

ERADICATION BASICS

If bed bugs are found in a unit of an apartment complex it's becoming common practice to also treat the adjacent units. We strongly recommend using inspections and monitors in the adjacent units first to save money. Eradication is costly because several factors make bed bugs difficult to eliminate:

- They build up resistance to pesticides. There are many ways they do this, and one of them is called *reduced cuticular penetration*. Basically the thick-skinned bugs, the ones that have a larger cuticle, survive and breed with each other, producing babies with thicker exoskeletons.
- The products that kill adult bed bugs typically aren't as effective on the eggs.

- Tenants who don't want to tell anybody they have bed bugs start self-treating with store-bought pesticides that drive the bugs deeper into cracks and crevices and into walls.

- Some tenants are hoarders whose stuff gives bed bugs too many nooks and crannies to hide in. Others pull furniture out of garbage bins or host down-on-their-luck friends or relatives who come with bed bugs.

- Some managers just don't want to see the problem. They think it's disgusting and refuse to arrange inspections.

Choosing the cheapest pest control company can also cause problems. To find the ones that are competent in bed-bug control, don't shop on price alone. The pest control industry as a whole is getting more skilled and quicker at bed-bug treatments, but they still take time to do properly. There is a difference between a company that spends thirty minutes performing chemical control and a company that spends two hours—and it typically costs more to hire the company that is going to spend two hours. Find out

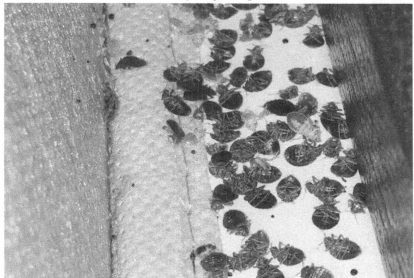

Dead bed bugs.

what those differences are, then get a per-unit price with a long warranty.

We get calls from properties that are continually getting bed bugs back in the same units. A hotel manager calls and says the bed bugs keep coming back in the same two rooms about every eighty days. He's hired a company to spray, and since its warranty is thirty days, he pays for them to come back. I'm thinking, *You have eighty hotel rooms, and you keep getting them back in the same two rooms? What are the odds that new guests are bringing new bed bugs into only the same two rooms? Your company is not getting rid of the problem. You need to make sure you're getting an adequate warranty. Thirty days is just not long enough.*

Our company offers a full-year warranty on single-family homes and six months for apartments and hotels. We want to make sure that it's long enough for you to know that the problem has been resolved. And if something comes up during that time, we can make sure it's taken care of. Sometimes it does take two or three months before the problem is resolved, especially when treating with chemicals. Six months should be plenty of time to confirm complete eradication.

TREATMENT BASICS

I have talked to managers who are paying for treatments but have no idea what the company is doing. You need to know what products and methods are being used and what follow-ups are planned. Let me repeat and reemphasize that: *You need to know what products and methods are being used and what follow-ups are planned.* Ask questions like, "About how long will it take you to treat a two-bedroom apartment? What products will you be using? How often will you follow up? How will you know that the treatment is successful?" If the answer is, "Treating a two-bedroom apartment should take us fifteen minutes," a red flag should go up. Both chemical and heat treatments take a while. I'll describe them both.

CHEMICAL TREATMENTS

Eliminating bed bugs with chemicals is possible, but it is difficult. A variety of insecticides are available in liquid and dust form. There are a few key things to know about the chemicals available today. First, the best results come from treating bed bugs directly. That means we need to find every last bug and egg—and that is difficult. Many of the products that work better on bed bugs do not work

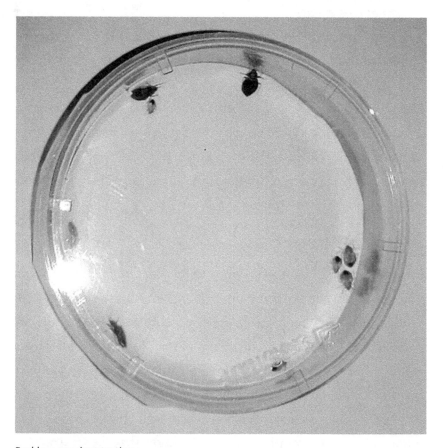

Bed bug product testing.

as well on eggs, while some of the products that work best on eggs don't have great results with the nymph and adult stages; therefore, using multiple products is advised. At the time of writing, we are seeing the best results on nymphs and adults with some of the newer dual-active products like Temprid and Tandem.

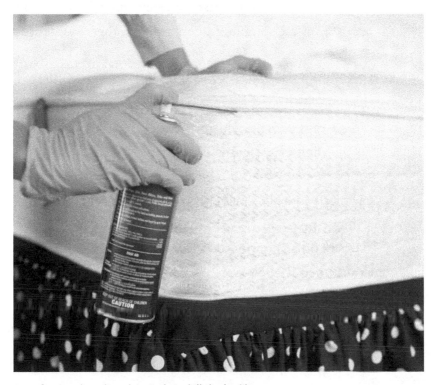

A professional applies chemicals to kill the bed bugs.

Desiccant dusts work very well with the silica gel-based products. We are continually testing products in-house to see what products work best, but we have found that different pesticides have different efficacies against different populations of bed bugs, meaning what works on some bed bugs doesn't work on others. Pest control companies should know which insecticides are generally more effective—but that's just the beginning.

What we started doing, and what we recommend other companies do, is collecting some bed bugs from your site and testing multiple products to find out what is most effective. Even inside one apartment complex, the bed bugs in unit B4 could be more susceptible to Temprid, while those in A14 are most easily killed by Tandem or Phantom. First we determine which product is most effective on each strain of bed bugs, and then we use it in the treatment.

It takes multiple chemical treatments, an average of three or four, to eradicate bed bugs. Chemical treatments must target bed bugs and their eggs directly, which often means using something more than the canister sprayer with the fan tip that a technician carries around. Some hand canisters have a crevice tool that helps get the chemicals into the cracks where bed bugs hide. The little baby bed bugs, called first instar nymphs, hide in the exuviae, or skin shed by other growing bed bugs. Vacuuming up exuviae underneath the bed is important because if you spray these sheds, the chemicals often won't penetrate and kill the hiding babies. Because of this and the need to reach cracks and crevices, these treatments can't be done in five or ten minutes. Technicians should be spending a good amount of time making sure that everything gets treated. We are getting faster and better at it, but it still takes time to find them all. A chemical treatment plan might also include using a commercial steamer along cracks and along the seams of soft items, especially when there are heavy infestations. If enough heat penetrates, it can quickly kill both eggs and adults, but steaming also takes time.

Whoever you hire needs to make a plan and tell you how they will treat electronics, kids' toys, and other sensitive items. We're not going to spray those toys down with pesticides, but we do find bed bugs in toys, televisions, and alarm clocks. It's possible to seal those items in a bag with a pesticide-impregnated plastic strip, which lets off a gas that kills the bed bugs but then dissipates when removed, leaving behind no residuals. Those strips work better for hard items like DVD players than for soft items like blankets, pillows, or clothing. Your pest control company should tell you how long the items need to stay in the bags, but I say the longer the better. A good treatment plan lets you know when all the bed bugs should be gone. Using heat for these items is our preferred method, as we have found that it is faster and more effective. Small heat chambers can be purchased or built, and these sensitive items, like some electronics and children's toys, can be heated to 130 degrees to kill any bed bugs and their eggs.

HEAT TREATMENTS

Heat treatments can be very effective at eradicating bed bugs, but not all companies do a great job with it. Bed bugs die almost instantly if the temperature reaches 122 degrees. So we try to heat the entire space to around 130 degrees and make sure the heat penetrates through everything, using electric or propane heaters and fans. The technicians have to go into the heated space and move items around and measure temperatures inside the couch, dresser, or mattress. In hotel rooms, we like to measure the temperature in the middle of the Bible, which is one of the last places to reach 130 degrees. The technicians are going in every thirty minutes or so to check temperatures and move things around.

It's not pleasant to work in those temperatures, but heat has several advantages. It is generally done in one treatment, includes everything inside that space, and afterwards, you can move on with life. You don't have to have everything you own sprayed down with pesticides. You don't have to treat all the kids' toys with pesticides. The disadvantage of using heat is the space is often a mess when finished because everything needs to be moved and rotated to make sure it is hot, and it is possible for some items to have heat damage. You have to move out guns, ammunition, aerosol cans, crayons, candles—anything that would typically be damaged if left in your car on a hot day. Everything that comes out has to be inspected to make sure not a single adult female bed bug comes back in. But even after removing

Bed bug heater.

the obvious heat-sensitive items, you may find something that didn't like the heat.

When you are picking a company to do heat treatments, make sure it has a way to measure the temperatures. I know that sounds obvious—but some companies are not measuring temperatures or not doing it frequently or carefully enough. Just putting a heater in a room for a few hours is not going to do the trick. Everything needs to get hot, including inside the walls and under the baseboards. We find that a direct-fire heater, which sucks air from outdoors through a fan past a gas-fired flame, can reach critical temperatures throughout the space in six to eight hours, contrasted with electric heat, which can take as long as twelve to twenty-four hours. These times vary depending on the size of the space being heated, how much stuff is inside, and how tricky it is to heat. My advice is to actually see one of their heat treatments. Walk through it with the technicians and see exactly what they are doing.

The heat has no residual effects on bed bugs once it's gone. So putting some pesticides in likely bed-bug harborages can help if, heaven forbid, a bug survives the heat treatment. Companies need to look for and address possible failure points in heat treatment. Some types of construction, such as cinder block, are harder to heat. If there is concern that heat is not reaching underneath the baseboard, you can always place pesticides down there.

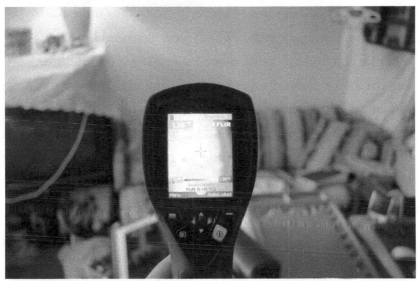

A professional checks the temperature during a bed bug heat treatment.

CLUELESS TREATMENTS

Sorry to say, there are some unscrupulous pest control people out there making bucks on bed bugs. They make my blood boil. They're treating bed bugs and don't have a clue what they're doing. How do you spot these? The more questions you ask, the better. Make sure they have a good concept of the things we've talked about in this chapter. You personally should see the bed bugs, or at least a picture of them at your property, before you authorize treatments.

Some companies give clients long instruction sheets that are almost impossible to comply with. They're asking the tenants to bag up everything they own, to vacuum, to treat all of their clothing. Preparations should not take weeks; they should be minimal.

Bug bombs are a big no-no because they drive bed bugs deeper into the cracks and crevices and disperse them to new areas, making treatment more difficult.

Sometimes we are called out to treat bed bugs and discover that there are none. Yes, some people and companies may lie about the presence bed bugs, for whatever reason—but more likely they are misidentifying insects. We've seen dermestid beetles, baby cockroaches, fly-fecal spotting, spider-fecal spotting, swallow bugs, and bat bugs misidentified as bed bugs. The management of a large row of more than one hundred townhomes called us for a second opinion about an infestation, and I saw what looked a lot like bed bugs but that were actually a little bit smaller, a little grayer. I saw them not only in the beds but also up on the walls. I walked out on the balcony and said, "Have you had problems with swallows?" Sure enough, there were swallow nests all over the sides of these buildings. The nests were covered in swallow bugs, which are also cimicids, in the same family of arthropods. They look like bed bugs, and they will bite people. They need to be controlled, but they don't require the extensive and costly treatment planned by the other company. That company insisted it was right, saying, "There's no other insect that looks like a bed bug." Wrong! The client almost spent more than $20,000 on unnecessary treatments in one hundred townhomes.

We also worked with a hotel that spent tens of thousands of dollars using bed-bug treatments for swallow bugs. They didn't need to. Swallow bugs and bat bugs are typically fairly easy to control.

You remove the nests (after the birds have migrated), and you do some perimeter treatments around windows and doors. Knowing what treatment your pest control company plans and comparing it to what you have learned in this book can save you from wasting money—and it can possibly protect your reputation.

I cannot cover everything there is to know about bed bugs in this chapter, but the biggest takeaway is to know what your company is doing for bed-bug prevention and treatment. Make sure they perform detailed inspections, have a well-thought-out bed-bug protocol, do good follow-ups, and provide documentation. Also, make sure you are seeing good results.

CHAPTER 5
BED-BUG DETECTION DOGS

When people call us with bed-bug problems, they want solutions fast. We need to be able to detect and eradicate bed bugs in the fastest and most affordable manner. After a lot of research and years of experience, we believe a vital part of this process involves what is known in the industry as a bed-bug detection canine . . . or what's known in my house as Radar, the beagle.

Radar, who has been working with me since 2009, was a rescue dog, abused as a puppy in Florida. The trainers from J&K Canine Academy near Gainesville found him in a shelter, and he really took to the program. In fact, when it's a long weekend and I don't work him very much, it drives him

Radar, the beagle.

bananas. The first day I was working him, I looked him in the eye—a mistake for a new handler—and he jumped up and bit me right in the forehead and drew blood. But now he's the biggest sweetheart ever. He's not always the friendliest with everybody else, but he's super loyal to me.

These detection dogs are working dogs, so they only eat when they find bed bugs. This is not cruel. Every day, my dog gets two cups of food. But he gets fed when he finds the bugs out in the field. And if at the end of the day he still has food left out of those two cups, then I have to plant bed bugs for him to find. I raise my own bed bugs, and hide one—secured in a sheer-covered vial—somewhere like a bed or couch. We search, he finds it, and he gets the rest of his food that day. He gets plenty of exercise working and doesn't go on a lot of walks. But these dogs absolutely love what they're doing. These dogs want a task.

HOW IT WORKS

A well-trained professional can be more effective at locating bed bugs, even at low infestation levels, with one of these dogs because they can sniff out bugs that are in places the professional cannot see, and they can search extensive areas quickly. Thoroughly searching large apartment complexes visually would be exhausting and extremely time consuming for a pest management professional. Using a dog allows us to handle those accounts, as well as universities and large hotels. Our company still chooses to do visual inspections alongside the canine inspections because we want to use every method we can to detect bed bugs. But using the dog allows us to speed up the visual inspections on these large searches. Even in a smaller space, it's an additional tool to be as effective as possible in detecting bed bugs. They're able to sniff out bed bugs where they're difficult to see, in wall voids and underneath baseboards. No method is perfect, but by deploying multiple methods, we get better results. We see companies get into trouble when they rely solely on one method for detection.

Radar searching for bed bugs.

Having a bed-bug dog promotes consumer confidence. Our best professional can complete a visual inspection and tell the client, "I cannot find any evidence of bed bugs here. Everything looks great," and many times the customer's response is, "Great. When can the dog come through to check?" They love and trust the dog to either prove or disprove that live bed bugs are there. It's just cool that a trained dog can sniff out a little bug that causes so much distress. It has been an amazing pleasure to work with Radar.

Bed-bug dogs are similar to those that sniff out bombs or drugs, in that they are trained to alert their handlers to the scent. This alert can involve sitting or scratching or pointing when it detects the scent. Studies have shown that these dogs can be very accurate, as high as 97.5 percent.[7] But not all dog and handler teams are the same, and a single team is not the same over time or in different environments. A lot of factors can affect the team's accuracy, including the time of day, whether or not the dog is sick or tired, whether or not the dog is distracted, and whether

7 Pfister, Koehler, and Pereira, *Journal of Economic Entomology* (2008), 13–96.

or not the handler is distracted or having a bad day. Air currents and sanitary conditions of the room can also be factors. It can be very distracting to a dog if there is food on the ground. The dog wants to eat the food and not search for bugs. Strong air currents can mask or pull the scent of bed bugs away from the dog's nose, making it more difficult for the dog to find the scent of live bed bugs. If the dog alerts in seven rooms, and upon visual inspection you don't find any bed bugs, the client will lose confidence. So the handler needs to make sure he is keeping up with the dog and that the dog is working well.

If I'm nervous or a little jumpy around an area, my dog can sense it. For example if I know another bed-bug dog alerted in an area, I may get nervous as we approach that area, Radar gives me a look that says, "Something's going on. What are you doing?" I need to make sure he is not going to alert falsely just because of the way I'm acting. Radar won't work with anyone but me, but I can work with our second dog, Webster, a beagle mix who goes home with a different member of our team. Every dog is different, and handlers and companies need to find out how each dog works best.

FALSE ALERTS

Radar and I have made some incredible finds and saved properties and homeowners thousands of dollars. But I've also learned that dogs are not always perfect. So we suggest that you get visual confirmation for all canine alerts. When a dog alerts, the handler should put the dog away in his crate or kennel, inspect the area, and be able to show you the live bed bugs. Bed-bug treatments are too expensive to deploy when you're not positive that you have bed bugs there.

At that point, you should have an immediate backup plan. We suggest that if the dog alerts and nothing is found, we monitor that area or come back and inspect again in a week or two to see if we can find anything. Since adult females lay two to three eggs per day, it's going to be easier for us to find bed bugs even in that short interval. What you don't want to do is ignore what could become a big problem later. If the dog handler is an expert and a pest control professional, he or she should be able to assess and discuss what the situation is and make recommendations.

SPECIAL SITUATIONS

Dogs are particularly useful in large, comprehensive sweeps. Visually inspecting two thousand rooms accurately is an incredibly daunting task for a technician, but we can accomplish this with one handler in a single month. This gives hotels, apartment complexes, and resorts a reasonably priced, proactive bed-bug detection method.

Hotel managers get a lot of complaints from guests, and they are not always sure whether to believe them. Having a canine and visual inspection really sets their minds at ease.

Visually inspecting empty apartments, especially after they've received fresh paint and new carpet, can be really challenging. But dogs can sniff out any bed bugs in wall voids, built-in wall units, and even adjacent apartments before the new tenant moves in. That can save property owners and managers a lot of money and hassle, and the new tenants avoid a potential nightmare.

Offices, schools, theaters, and public transportation can be really difficult sites to inspect visually. These complex environments give bed bugs many places to hide. So using a canine in these nonbedroom sites really helps. In a home, we use our bed-bug dog to determine which rooms need full treatment. We often don't have to heat the entire space or structure. That can save a lot of money.

OVERCOMING SKEPTICISM

We were called to an eighty-unit apartment complex that had been struggling with bed bugs for a while. We put together a program and talked with the management about it. They wanted us to go through and inspect all of their units. They gave me keys, and said, "Here, have at it." Radar and I went through all eighty units, finding twenty-one apartments that had live bed bugs. But when we got done and reported our findings, I got a disturbingly skeptical phone call from the vice president of the company. She said that they had just had another company inspect all the units the week before, and they found no bed bugs. She was fairly adamant that I needed to explain myself.

"Yeah, both of us cannot be right," I admitted. "I'm saying there are twenty-one and they're saying there's none. One of us has to be wrong. I've got pictures of every single one of these, and I can send them to you and document exactly where they were found and how they were found and what color couch they were found on."

You see, every time that Radar and I find live bed bugs, we try to take a picture of them to document that bugs were found. So every time I found them at this complex, I took down the number of the apartment and snapped pictures. It gave a lot of credibility to us and is a good example of the importance of documenting findings. Because I had the documentation, we won the account. The property still has an occasional bed-bug infestation brought in here and there, but these are few and far between, and we find them quickly and eliminate the problem.

HORROR STORIES

One evening, we had a call from a hotel with a frantic request for heat treatments in numerous rooms that very night. And our first question was, "What makes you think that you have bed bugs in all these rooms?" The answer was that a dog had found bed bugs in eight of twenty-five rooms searched already, and they were scared how many more rooms might be involved.

"Slow down," I said. After confirming with them that the other company had not visually confirmed any of the canine alerts, I said: "Let's not heat tonight." I persuaded them to let me go through that night, about ten p.m., to search the eight rooms with Radar—and we did not find bed bugs in a single one. When we lifted the beds, we found discarded McDonald's wrappers, beer bottles—everything but live bed bugs. Maybe the other company's dogs were alerting to that food, maybe they were just having an off day. But visual confirmation saved them a lot of hassle. We monitored those rooms again two months later, and there were still no bed bugs.

We had a similar situation at an apartment complex where another company's canine team had alerted to bed bugs in fifty-two units, and our inspections, which included visual and canine, found that not a single one had bed bugs then or anytime since. We also got called out to do heat treatments at sixteen military barracks in which there were canine alerts. In that case, we had to convince a woman who said, "Dogs can never be wrong," that another inspection might save some money. Again, we went through all sixteen barracks, inspecting visually and with a canine, and no live bed bugs were found, then or since. They saved not only $30,000 in expenses to heat the barracks but also the negative publicity that they were worried about.

Several months after I started working with Radar, I was called to a large three-story office building where managers said they had bed bugs for several years on the third floor and treatments weren't working. Radar started searching, and about halfway through a row of cubicles, Radar started alerting in each one. I thought, *Holy cow. We've got massive bed-bug problems in this area.* But after I put Radar away in his crate, I got on my hands and knees, searched those cubicles, and couldn't find any sign of bed bugs.

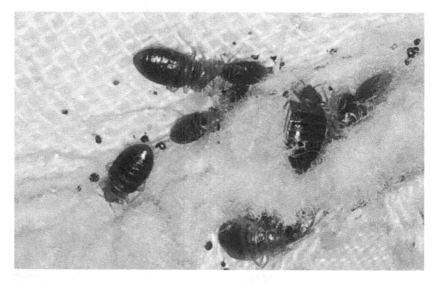

Bed-bug infestation.

The employees started showing me glue boards, and sure enough, there were cimicids about the same size as bed bugs but a little grayer. Bed bugs are a type of cimicid, but we don't typically find them on sticky traps and not all the same size like these bugs. Because other companies had identified them as bed bugs, my new client insisted that they must be. So we sent the specimens to the entomologist at the state extension office. In the meantime, I looked at the outside of the building and noticed that there were a few storm drain covers missing and a lot of bird droppings on the roof. So we put a hose down one of the storm drains to see what would come out. Sure enough, two big barn owls and the carcass of a barn owl and thousands of bugs came flushing out. It was pretty disgusting. Bugs were feeding on barn owls roosting in this drain, and as the population overgrew that, they found a way through the ceiling onto this top floor, where they could feed on people. But as the state entomologist confirmed, they were not bed bugs but Mexican chicken bugs. So instead of wasting money on years of bed-bug treatments, all that really had to be done was to get rid of the barn owls and replace the storm-drain covers. The lesson is that it's critical to properly identify the insect.

Bed bugs in the corner of a box spring.

I don't want to dissuade anyone from using bed-bug detection dogs. We think they are an incredible tool, but I want to stress the importance of getting a visual confirmation. If you see the bugs, then you don't have any concern with wasting your money. I know some people worry that you can't find them visually. You can. It takes some effort and digging around with a flashlight, but in almost all cases, we're able to show you a live bug. And if we can't, then let us go back in a week and check it out again.

CHAPTER 6
DON'T BE THIS CUSTOMER

Entomologist Jeff White is a national expert on bed bugs who has been featured on Animal Planet's *Infested* TV series. White, who also hosts a video series called Bed Bug TV on BedBugCentral.com, used Twitter to tell his fans about how he was thrown off the property of a hotel where he was staying.

"Guess who had #bedbugs in their hotel room last weekend? guess who was told they were lying and 2 get off the property? #thisguy," tweeted White, @JWhiteBBTV, on November 1, 2013.

That's an example of the attitude some clients have that impedes good pest management and causes them to spend more money and have more stress, hassle, and loss of reputation. Please don't be one of these customers.

DENIAL AND LACK OF CONCERN

We have seen public libraries where people are regularly finding bed bugs—in chairs, couches, or books—and administrators don't want to do anything about it. They shrug their shoulders and move on. We have been to a food-processing plant that was willing to keep selling products with severe infestations of Indian meal moths and mice and not willing to address the problems. We've had grocery stores say no to inspections aimed at finding the source of the mice and German cockroaches rampant in their back rooms and food-prep areas. We have seen restaurants with German-cockroach infestations so bad that the piles of dead cockroaches were over an inch thick, and they don't do anything about it. They may be reluctant to spend the money on treatment, or they may be hoping the problem will go away by itself, which it won't. These are serious problems that need to be taken care of.

FEAR AND LOATHING

We meet a lot of apartment managers who don't want to go inside the units to see the cockroaches or other problems we want to show them. I know that it's icky to many people. Pests are among Americans' most common phobias. But I really feel it's important that managers of properties in any industry have an idea of what's going on and be at least a little bit involved. The pest control operator will do the majority of the work, but you should see the problem yourself because that will motivate you to act. You need to see the cockroaches that people are eating or living with. You need to see the mice droppings. If you really can't face the pests in person, at least ask your professional if they can take pictures.

We have clients who think that if they walk into a unit with cockroaches or bed bugs, they're going to take them home. I am around bed bugs all day long, and I have never taken one home. Wise precautions include not sitting down on couches or beds, not leaning against walls, and being careful about what you set on the floor. While treating cockroaches at restaurants or apartments, we have met up with maintenance people in full hazmat suits, looking like they are going into a nuclear disaster, but that is not generally necessary. Usually, some tape around the pants legs to keep cockroaches from crawling up is all you need—and that is when the infestation is severe.

Beyond fear of the insects themselves, there's fear about the money that's going to be involved. A lot of problems get ignored because a manager doesn't want to go to the boss and say, "We have these issues. We need money to solve that." Both apartment managers and tenants are also terrified of finding problems because neither wants to have to tell the other that they have a problem.

COMMODITIZATION

Thinking that all pest control companies are alike is another attitude we deal with a lot. Pest control is a technical field, and it's not simply walking around with a hand canister and spraying baseboards and throwing a few traps down. A quality pest control company knows the best products to use for each insect and each case. Technicians inspect the properties. They can ID pests. They can find conducive environments and either correct them or explain how the restaurant, hotel, or other property can fix them. The best companies see the whole picture and solve the problem.

It should be obvious that not all pest control companies will hire people with the same level of training and professionalism. Can they tell the difference between a carpenter ant and a field ant? Can they ID beetles in a food-processing plant? Do they know how to pest-proof a door? Can they communicate that to a client? Can they calm down a tenant distressed over cockroaches by explaining what's going to happen?

FALSE EXPERTISE

Some managers or owners who have read an article on the Internet think they're expert on the pests we're dealing with. We're told how to do our job, and sometimes what they think is just completely wrong.

Restaurant managers believe that nothing can be done to prevent flies. (The many treatment options were described in chapter 3.) Apartment managers say, "Just throw the bed away and you get rid of all the bed bugs." Almost always bed bugs will still be in nightstands, baseboards, or walls. We have clients and managers who say, "Just put some diatomaceous earth down." That desiccant dust helps control bed bugs, but we've seen homes completely white with diatomaceous earth that still had live bed bugs in the cracks and crevices. We hear from clients that bed bugs never spread from unit to unit, but that's a very real possibility, whether the pests are traveling through wall voids or on people or products. We hear that you can't treat spiders using liquid pesticides, which is not true.

REFUSAL TO DO THEIR PART

Pest control is not just spraying pesticides. We have to address the triangle of life: food, shelter, and water. If we can eliminate a couple of those, that really helps. So we are asking managers to cut weeds, fix water leaks, and pick up food scraps. To find these conducive conditions, we ask to search adjacent units, or in food-processing plants and restaurants, to have an inspection aisle against the walls. But when we bring up inspection and prevention, a lot of time we run into the attitude, "You're the pest control company, so you need to figure it out and fix it."

Customers should understand we are not making excuses when we point out conducive conditions. A company that wants to help you correct these issues is trying to give you better results. It's going to save you money and headaches. It was established by Francesco Redi in 1618 that pests don't spontaneously generate.

There's a reason why the pest is there, and our job as pest control professionals is to find out why they're there and how to fix the problem. So participate and be a part of the solution.

UNREALISTIC EXPECTATIONS

I wish I had a magic wand and could wave it around all of my clients' facilities so all of the pest problems would go away. There are times when it takes extensive problem solving, not a fifty-dollar spray job. An infestation that has become worse over months and years often cannot be solved overnight. This goes the other way as well. We have met clients who believe that their problems can never be solved or will take years to fix. That's too long for most pest issues. I don't want to live in an apartment or eat at a restaurant that can't control its cockroaches for years. Never having another fly on your property is probably not going to happen. But there are some great things we can do to help control and to monitor and make sure that issues get taken care of.

A landlord cleaning out a vacated apartment discovered German cockroaches, and he called us asking us to make sure they were all gone by the next day, when a new tenant was scheduled to move in. Not only did the manager want all the cockroaches dead in twelve hours, but he insisted that the new tenant not see a single dead cockroach. We walked into the apartment, and there were thousands of cockroaches. We offered a treatment timeline, which was fast but did not involve solving the problem in one day. Some clients don't love to hear that.

"NOTHING WORKS"

An apartment manager told us that nobody can get rid of cock-roaches, really, so he might as well find the cheapest company to perform the minimally required service.

If you're not doing what needs to be done, the problem will only get worse, and you're going to end up spending more. If your entire apartment complex is littered with cockroaches, you'll pay for it in your online reputation and decreasing occupancy rate.

UNWILLING TO PAY WHAT IT TAKES

A large grocery store in a poorly sealed, poorly maintained building had mice, cockroaches, and major fly issues. And the budget for pest control was fifty dollars a month.

That wasn't going to be enough. We try to work with a client's budget as best we can. But there is a minimum required investment on some of these pest issues. Many of these clients want a quick fix, but what is needed is a "surgery" that they are unwilling to fund. Quality pest control is an investment in client and guest happiness: it is money well spent.

There is a difference between a $50 cockroach-treatment program and a $200 cockroach-treatment program, and the dif-ference is in time spent and products used. It's not a case of the $200 company being greedy. It may be what is required for that situation. Other situations may require more than $200 and some less than $200. A good pest control company that wants to help you will inspect and assess the situation and tell you what is going to be necessary to accomplish the job.

CHAPTER 7
RESEARCH AND DEVELOPMENT

B efore we began servicing a certain hotel, they struggled with bed bugs for years, experiencing eight to twelve infestations a year, never in the same room but scattered at random around the hotel. We were inspecting the entire hotel quarterly, and this was helping us to locate and eliminate infestations quickly. But there were still more bed-bug infestations than you want to see in a hotel, and we wanted to solve that problem. This was in 2009, before preventive bed-bug treatments were really accepted by the pest control industry.

We did some research in-house with desiccant dust and had seen pretty good results on bed bugs. We were starting to use it at customer sites, but the hotel management was reluctant to try it. Finally, we persuaded them to let us treat all the rooms proactively, to make every room a hostile environment for bed bugs. We thought the dust might keep bed bug infestations from taking hold and spreading—or at a minimum, keep the population down so they would be easier to eradicate when we did find them. But for the next two years, the hotel didn't have a single bed-bug infestation. In the years since, we've done these preventive treatments in thousands of rooms of hotels, courthouses, firehouses, dormitories, and apartment complexes with similar results. It doesn't prevent 100 percent of the bed bugs, but it really knocks down the problem.

As of 2015, most companies were still not performing a similar preventive service, but it was starting to be recognized as a best practice. Dr. Michael Potter of the University of Kentucky published a research paper in *Pest Control Technology* magazine in August 2014 on this practice of using desiccant dust, specifically silica gel, for bed-bug prevention. The next year, he presented his findings at a conference I attended and asked everybody to raise their hands if they were doing the preventive service. Of a group of 200 to 250 pest control professionals in the room, no more than five of us were doing so. He just looked at us and said, "This is going to be the future."

We started it in 2009 because we were paying attention to what was out there and doing research and development in our office. Because we kept studying and learning, our clients saved a lot of money on eradication treatments. To me, pest control is an exciting industry because it always has new research, products, technologies, and treatment methods. It moves so fast that you can always tell immediately if a company has kept up with the research.

A CULTURE OF INNOVATION

Finding new ways to be more effective will deliver better results with less impact on people and the environment. Pest control professionals who are constantly studying and learning can better diagnose unusual problems, better identify insects, and be able to match products and techniques to the client's needs. They are also more likely to be able to confidently explain what they are going to do and why, what to expect, and how to follow up.

At Thorn Pest Solutions, feeling like we are on the cusp of the industry and doing all we can to take care of clients keeps employees happy and creates an exciting, invigorating company. The more my professionals feel like experts, the more they enjoy their work. It is more rewarding to diagnose and solve a problem than continually treat it—like putting on a Band-Aid.

My dad was a dentist, and growing up, I always saw him studying and learning. I never saw my dad sit in front of the TV, at least not without a book to read. I picked up his habit of continuous learning, and it helped me as a premed student at Brigham Young University and as an entrepreneur running a company.

CONTINUING EDUCATION

Every state requires some level of certification for pest control work. The state of Utah requires us to take a series of exams to become a licensed pest control operator. Then you have to complete twenty-four continuing-education units—roughly a classroom hour per unit—every three years. We take that as a bare minimum in our company. We generally try to have between sixty and ninety hours of continuing education per year, which is on par with doctors and lawyers.

I became an ACE entomologist by passing a major exam, and I maintain certification from the Entomological Society of America by attending regular classes. The NPMA provides a company-level certification, QualityPro, for which all of my technicians have to pass an exam. To keep that designation at Thorn Pest Solutions, we sign affidavits and have periodic audits to ensure adherence to the many QualityPro guidelines, which include:

- criminal background and motor vehicle records checks on those we hire

- drug-free workplace policies
- minimum insurance requirements
- service vehicle standards
- dress policy standards
- ensuring that employees are tested and trained to the highest industry standards
- on-the-job safety policy
- proper pesticide handling
- termite warranties and service agreements
- customer communication policy

The NPMA provides education through newsletters, books, training programs, and the huge annual Pest World Conference. The Entomological Society of America also offers conferences and scientific journals where researchers publish their work. There are many other great professional resources, including Gary Bennett's *Truman's Scientific Guide to Pest Management Operations*; industry magazines, such as *Pest Control Technology* and *Pest Management Professional*; and state agriculture department extension office websites.

When I first bought my company, we didn't own a single book. Now we've got an entire library. The conferences opened up my eyes to how much there was to learn.

The first one got me interested in bed-bug detection dogs. More recently, a class by Dini Miller of Virginia Tech sent me back to the drawing board on cockroach treatments, and within two months, we scrapped most of our chemical spraying and began using baiting at apartment complexes.

One of the things Dr. Miller talked about, based on field research using monitors and bait in Virginia public housing, was that we're always worried about how clean the tenants are—and she found that it doesn't matter. Cockroach baits, if placed sufficiently and in the right places, are successful. Our company wasn't using a lot of baits at the time, and when we did, they were not working very well. This really taught us a lot about how to use them. Baiting programs have some big advantages, and my favorite is that the tenants have to provide little to no prep or cooperation.

So after seeing this presentation, we went back home to Utah and immediately began reading more about baiting programs. We tested products on cockroaches in our office lab and then tried it on-site at a few apartment complexes. We quickly moved most of our cockroach programs to baiting, and it has been very successful.

PERSONAL RESEARCH

I have learned a lot about bed bugs in the past ten years but still want to understand more about their behavior. For research, and for our canine program, I breed my own bed bugs by the thousands, which means feeding them what they eat, human blood. It took me some time to figure out the best way—which is to put a jar containing the bugs, with a lid of sheer material, on my arm or on my belly, and let them feed. The first time I did that in bed, my wife was a little weirded out. It was prickly, itchy and unpleasant to have five hundred bugs feeding on me, but I've since become immune. Not everybody's allergic to

Bed-bug jar with live bed bugs.

bed-bug bites. Some people get itchy bumps right away; others have a delayed reaction or none at all.

There will be more research on products that are effective but have lower impact on the environment. Putting a new pesticide with a new active ingredient on the market, with all the testing the Environmental Protection Agency requires, is incredibly expensive, perhaps costing $100 million. So a lot of money is at stake in research. But so is something more basic. The innovations in pest management parallel with an increase in life expectancy. A child born in 1900 had a life expectancy of forty-nine years, while a child born in 2010 had a life expectancy of seventy-eight. Public health officials attribute the quality of life we have today to three things: better pharmaceuticals and vaccines, better sanitation, and better pest control.

The proof is in history and the world around us: rat fleas spreading plague and killing a third of Europe's population in the fourteenth century, lice spreading typhus that wiped out most of Napoleon's army when it invaded Russia in 1812, and today, every thirty seconds, a child in Africa dies of malaria that's spread by mosquitoes.[8] Rats bite forty-five thousand Americans every year.[9] According to the American College of Allergy, Asthma, and Immunology, two million Americans are allergic to stings from bees and other stinging insects. Half a million Americans yearly visit emergency rooms due to reactions of being stung. Pests spread all sorts of diseases, including typhoid, cholera, salmonella, and dysentery. So it's vitally important that we continually research, study, and offer better results.

---

8 Frederick M. Fishel, "Pest Management and Pesticides: A Historical Perspective," University of Florida IFAS Extension, https://edis.ifas.ufl.edu/pdffiles/PI/PI21900.pdf.

9 "Pest Management Industry Fact Sheet," NPMA, http://www.npmapestworld.org/default/assets/File/newsroom/IndustryFactSheet2013Curlnumbers.pdf.

CHAPTER 8

MEANWHILE, BACK AT THE OFFICE

One of our clients, a food-processing plant, was dealing with significant rodent issues before they hired us to investigate and fix the problem. After inspecting the facility, we drew up a plan that included placing rodent bait stations outside and monitors for mice inside. As we began to service this account, our professional physically checked all of the monitors and bait stations. His regular rounds involved scanning each device's bar code and recording his activity and observations on his iPad. He also unlocked the exterior stations and replaced the bait every month.

There is a field behind the plant, so we assumed the mice would be active in that area, but after a couple of months, one of our office team members checked the records and noticed that all of the bait stations in front of the facility were experiencing rodent activity while there was little activity out back where we had placed more of our bait stations. The next time the professional went out, the office staff team member advised him in writing to increase monitoring on the front side of the building near the entrance. The tip paid off—the professional followed up on the tip and found that a little stream with low-lying vegetation all the way across a parking lot was harboring a lot of mice.

Pests always appear for a reason, but it's not always easy to see. In this case, the office staff team member took the initiative to analyze the reports and, sure enough, she saw something that our professional in the field hadn't seen.

WHAT SETS US APART

Up until this point, I have talked about best practices in the pest control industry as a whole and not our company specifically. In this chapter, I will cover some of the things that make our office unique and helpful. Although I believe the topics covered in this chapter would be helpful for other companies, I acknowledge that there are many successful ways to run an office, and our way is not the only way.

A friendly and helpful office gives our clients a great experience and keeps them happy with us and with their services. They are not dealing with a grump but someone who is genuinely happy and just there to help. We're in business to help, and that starts with a helpful office team. They do all the tracking, follow-up, billing, and booking of the appointments. We have the ability to track when appointments need to happen so that when a professional tells a client, "Hey, we need to follow up on this ant job in two weeks," it gets scheduled, and someone shows up. Technicians can do the best work possible out in the field, but if there are failures on the back end in the office, the company doesn't look professional. The office team should anticipate the needs of the clients. And to do this, they have to listen and care.

We have more walk-ins than you would expect. Many pest control companies are in industrial areas, but we're on State Street, the big commercial thoroughfare in Utah, and have a big sign. So customers come in to pay their bill, ask questions, drop off specimens for us to examine, or just to say hi to us. We also offer classes for restaurant, hospitality, and food-processing workers, either in our offices or their locations, whichever is more convenient.

The service is all about the client, and that should be the focus. We care more about the clients than we do about money. We strive to develop this attitude in everybody in the company, so they greet each client warmly. They smile and they listen. Some people have this ability innately, and those are the people that we try to hire for our office. But this is also a talent that can be taught, so it is part of our training. Because in the end, they have to get that they are there to serve the clients, or they are not going to fit in with our team.

At Thorn Pest Solutions, we make every effort possible to make sure our customers are satisfied. Regularly, the leadership group—myself, the head of operations, the head of administration, and sometimes a representative from sales—meets weekly and gets a report from our client-happiness representative, who represents the clients and tells us how they are feeling. We analyze this report in order to identify what we're doing well and how we can serve our customers even better.

CROSS-TRAINING

Cross-training the office staff on the technical aspects of pest control is critical to them being able to help professionals out in the field know what's going on. If a billing person is the only one available to answer the phone, she can ask the clients the right questions and talk knowledgeably. People call in and say something like, "I've called fifteen companies, and none of them can tell me how they are going to treat for ants." Customers commonly hear things like, "I don't know how, but I'm sure our company will take care of it."

They respond to us positively, saying, "You guys were the first ones who asked the right questions, told me exactly what the process was, and what I should expect." Not having answers is not very helpful. All companies should be able to explain their services—and this includes the office staff, not just the professionals out in the field. Before you spend your money, you need to have an idea of what the company is going to do and what you're going to be getting for your money. You should not have to assume or guess that we as a company know what we're doing. You should be able to hear what we're going to be doing.

I know that our office team does not know everything about pest control, but I expect and train them to have a basic knowledge—to be able to listen and explain why we do what we do. Our office staff attends all the same training as our professionals, except for the hands-on field training.

TRACKING OUR SUCCESS

An up-to-date pest control office tracks and displays numbers so the whole team knows how well it is performing. We have a couple of big screens displaying a dashboard full of information. Any time we get a complaint, it goes up for all of us to see. We also display compliments, how many appointments were booked that day and that month, the revenue for the month, how many phone calls were made—a full report card. It gets updated every day, and the whole company knows exactly whether we're on track or not.

One dashboard has all the professionals' data, such as how much time on average they are spending on an account. We track that to make sure they are spending *enough* time, not because we want them to spend less time. The professionals can see their callback ratio to know how often they get called back to accounts after treatment. We have benchmarks that we set and want everybody to reach for, and we offer support to help get them there. We email our professionals a weekly accounting with all of the data crunched. If you're a professional, you need to know: What am I doing well? What do I need to work on?

Using data this way is not that common yet in the pest control industry, so instead of buying a ready-made software package for some of this, we had to painstakingly put together our own spreadsheets. We're constantly adjusting our dashboards to make sure the tracking is helpful, not burdensome.

UP-TO-DATE COMPUTERS AND SOFTWARE

Computers are affordable enough these days, so there's no excuse for not having good, fast computers running robust software that can track everything that's happening. Our software has the ability to take in a lot of data. The more data we can collect, the more reports the software's able to run. For example, we run trend reports on our bait stations, which are each identified by a unique bar code.

If a location has fifteen exterior bait stations, we can run a report for twelve months and see exactly which ones have gotten activity. That may result in our adjusting our treatment plan. Say a client has light traps inside a building for flies. When we change out the sticky boards once a month, we count how many flies are on those boards, and we put that data into our system. A fly light may have fifty-two flies this month and sixty-four flies next month. Those numbers alone have little meaning, but if we collect that data for months and years, we can tell whether our fly-control services are working, and we can show a client exactly what happened.

GPS tracking data from our vehicles tells exactly where each professional is at any given moment, how fast they're going, or whether they are parked. This information allows us to deploy them quicker to client sites and also helps with quality assurance. If a service call should have lasted about an hour, but the vehicle was only there for twenty minutes, we will see that kind of big disparity and find out what happened. Were we not able to service everything? If we ever have issues, we can go back in history and check those things.

Our software doesn't just make reports for our team to see and analyze, but it also feeds into an online dashboard. Our clients can log in from our website, and they can pull any report themselves if they want to. Or they can call us, and we can run it. Our software lets us track every phone call with a client and store notes about what was talked about and what was done. So if they ever call back with an issue, we can always see if we discussed it and what was said.

TRAINING FACILITIES

Having a place to do hands-on training helps develop not only technical skills but also customer service. An up-to-date pest control office needs a place for the team to train and to talk about how things are going. A computer display that everyone can see at once is very helpful.

Another technical tool for training is the ability to record phone calls and listen in. It comes in handy for quality assurance—to be able to look up any phone call and see what was said and what we could have done better.

CUSTOMER SATISFACTION

Does doing everything we have discussed above mean that every one of our clients is always happy and completely satisfied? No. We aren't perfect. But we use the following methods to help remedy any problems that we can:

- **Focus on the client, not the money.** We need to think about your needs and try to see the situation from your perspective. Sometimes we need to drop the professional attitude and have a regular, personal conversation, in which we stop talking and just listen. This shows that we care much more about how the client is feeling than the money. If we have to lose money on some jobs, that's fine. We can't do that all the time, but we need to do whatever we can to focus on taking care of the client.

- **Get out in the field.** There's more on this in the next chapter, but the essential point is that managers especially need to get out in the field so they can focus on the client. Managers can get too overwhelmed with the numbers and the operations and everything else that they have to do. But they can't lose sight of the customer.

- **Have amazing follow-up.** We need to track and do follow-up on everything that we do. We have to know and ensure that we're getting the results we wanted. And some of that is done proactively, meaning that follow-ups are automatic when we treat for rodents or cockroaches. Sometimes that's as simple as a phone call: "And how are things going with that ant work? It's been a week and a half. Are things better?"

- **Do what is not expected.** We go the extra 10 percent and do what the customer was not expecting. We provide extra follow-ups, send them cards, and perform an extra level of service that is not spelled out or expected of a pest control company.

- **Anticipate a client's needs.** If we see a need, we should not hesitate to help. This might be a phone call, a visit, or a service that is proactive. This kind of great service goes a long way toward the client having a great experience.

CHAPTER 9
GETTING OUT INTO THE FIELD

I mentioned in the last chapter that upper management needs to get out of the office and into the field sometimes to focus on the client. When we are on-site, we know what pest control looks like, smells like, and tastes like. Tastes? Yes.

One time I was sitting in the truck eating a chocolate doughnut during a break on a job and decided to lick the icing off my hand. That's when I found out what a black widow tastes like. You see, I had smashed a black widow spider under my hand a little earlier and didn't notice the creamy black residue. Let's just say it did not taste like chocolate.

When I am in the office making a service plan for a commercial account, I am like an architect drawing up blueprints for a home. As detailed and expert as those blueprints might be, the architect is likely to encounter surprises at the construction site while checking on the progress before the house is completed. So I'll show up on-site occasionally, probably find things that aren't done exactly as I specified or intended, and tweak the plan. Seeing things in three dimensions and changing the plan accordingly, I am much more likely to get the end product I want: a client satisfied with the result.

TOP MANAGERS IN THE FIELD

Especially for tough jobs, the boss should get out from behind the desk to make sure our vision is being carried out. That starts with focusing on the client experience.

As a company and its management grow, they often begin to focus on the growth itself—the numbers and operational challenges. They leave it to the frontline team members to handle the client experience. Getting out in the field shows our team members that we care about both them and that client experience.

Speaking directly to clients and the frontline team members out in the field at least once a week gives managers insight into what the company can be doing better.

We can see the results of our work, analyze what went well in the service, and note what skills we might be lacking. In this way, we can design and focus our training. Are we getting the results that we and our clients want? A pest control company must be results driven. So no matter how much our company grows, my leadership team and I will be getting out into the field to get the true story. We're in business to help our clients, and that's where our focus should be.

Getting out in the field also helps with quality control. Is what is happening out there what we want and what we promised our clients? Are we treating clients the way they should be treated? Are we documenting everything the way we should so that the results are easy to see? Are we doing what we said we would do? Are the clients thrilled with our work? These questions need to be asked, and the answers or findings should be reported to the rest of the company. The best way we can grow as a company, not just in revenue but in quality of service, is by constantly improving.

We must understand our biggest weaknesses so that we know what needs fixing. We should be performing quality-control audits, getting out in the field throughout the year, and collecting data on a weekly basis. We should also be meeting together and talking about what we found out—what's going well, what's not going well, looking for patterns, and trying to spot any shortcomings. This needs to be every single week, not just once a year, so that we can pivot and change when we need to. If we relied on annual audits of our accounts, it would not be nearly as beneficial for our clients.

APPLYING RESEARCH

What happens in the lab is not always what happens in the field. Therefore, when we do research in our in-house lab, we should take what we learn and try it out in the real world. We should be tracking the field results just as much as the lab results in order to validate the research.

For example, we've researched different types of baits by trying out new products on the live German cockroaches that we keep in our lab. By seeing which baits the insects are taking to, we have found a lot of cockroach baits that do the job very well. But in one particular case, for example, a bait passed the test in the lab, but we didn't get the results that we wanted on-site. There could be several reasons, but probably the main one is the cockroaches had already developed an aversion to that type of bait because it had been used too continuously on the property. So we had to change the products we were using and change how we applied them to get the results we needed.

If we are called to a new property that we know has had repeated treatments from other companies, we'll often do a field experiment. We take several different units—apartments, if it's a multifamily housing project—and treat them differently to track which baits are working best. We can later apply that product to the entire complex. After that, we can rotate baits to make sure we're not producing bait aversion.

I don't want to give the impression that we have a huge in-house lab and spend all of our time doing incredible research. It's on a small scale, but we raise bed bugs, German cockroaches, and flies, and we test various new products as we read about them. Cockroaches, flies, bed bugs, and mosquitoes, with their brief reproductive cycles, all gain resistance to insecticides very quickly. This resistance can vary from region to region, inside a city—or even inside a single building. So regularly testing products in the field is important.

ASSESSING RESULTS

Knowing whether or not we are getting good results can be tricky sometimes. We have to look at several different factors to get a sense of how things are going. The first one will always be client reactions. Do the clients smile when we walk in? Do they have good body language? Are they happy to see us? Do they compliment us? Those types of reactions will give us a good sense of how things are going for that facility manager.

The professionals on-site should also be asking the clients questions about how things are going. What frustrations have they had? Are they seeing the results they want? Talking to them about what they've seen, what they're experiencing, and whether they're getting complaints is important to see how things are going.

And the pest control company managers should be asking similar questions of the professionals on-site.

Even if the clients and professionals on-site are pretty upbeat and positive, inspections are the best way to see if we're getting the results we want. To gauge, we have to get on our hands and knees and look around, and monitors are a great way to catch things that we're not able to see in the physical inspections. Combining results from the inspections and the monitors into spreadsheets or graphs, or both, can give us a very visual look at how things are going. That's a great way to look at trends and see hotspots in a specific part of a property.

Not everyone in the company is going to perform pest management the same way I do, but by getting out into the field, I can make sure that the key components are there and that clients are being well taken care of.

CONCLUSION

BE PROACTIVE WITH EXPERT HELP

I want managers and owners of apartments, restaurants, hotels, schools, food-processing plants, and hospitals to make great decisions, save themselves money, and experience the best results. Pest problems are costing many of these clients tens if not hundreds of thousands of dollars a year, and they need an expert to help them—but the problem is they don't know how to find one. I want to help. A lot is at risk beyond money when pest problems are ignored, misdiagnosed, or treated with anything short of total determination to eradicate the problem. Pests spread disease, harm business reputations, and make life unpleasant.

My professional life and my company, Thorn Pest Solutions, have been devoted to providing the best possible service by constantly learning and improving, listening, and communicating well. It all comes down to helping people and improving their lives.

When clients do business with us, they benefit from the core values that inform our daily operations and interactions. These core values are what drive our work and have helped us become one of the most trusted businesses in our field.

OUR CORE VALUES

1. **Appreciate, Educate, and Wow Customers.** We are in business to improve the quality of life for our clients. We dedicate ourselves to go above and beyond in everything we do.

2. **Constantly Learn and Improve.** We are only as good as what we know, and that is why we spend hours in training meetings, seminars, and lectures learning all that we can. We are always working to improve everything we do.

3. **Offer Great Value.** We strive to offer you the best bang for your buck. We want every penny spent with our company to be well worth it.

4. **Be Innovative and Take Risks.** Our company is constantly innovating new services. We were the first company in Utah with a bed-bug dog, bed-bug heat treatments, organic-based fertilizers, and season-long wasp-control treatments.

5. **Communicate Clearly.** We have invested heavily in technology to plan, track, and schedule your services. We know good communication and follow-through are important to bringing you the best in pest management.

6. **Be Passionate and Positive.** We love our work, and we hope you can tell by our eagerness to get things done right. This is our life's work, and we are proud to serve you every day!

GLOSSARY

Alate: reproductive ant or termite that is winged

Alitrunk: thorax of an ant

Arthropod: an invertebrate animal that belongs to the phylum Arthropoda, including insects, spiders, and crustaceans

Bait: an active ingredient placed in food or other attractive materials to get the pests to feed on them

Cimicids: the family of small parasitic insects, including bed bugs, that feed on the blood of humans or other animals.

Commensal rodents: rodents that try to live with or around humans (examples are the house mouse and the Norway rat)

Complex metamorphosis: the four-stage process (egg, larva, pupa, and adult) some insects go through to develop into adults

Cryonite machine: a piece of equipment that shoots carbon dioxide to rapidly freeze insects

DDT: a synthetic pesticide widely used in the 1940s, 1950s, and 1960s that was banned in 1972 by the US Environmental Protection Agency because of concerns about adverse effects to wildlife and human health

Desiccant dust: any insecticide in dust form with active ingredients, such as silica gel or diatomaceous earth, that dry out and kill bugs

Exuviae: the shed skin of an insect

FIFO: first in, first out, as in a pattern of managing inventory

Gaster: the abdomen of an ant

Instar: the juvenile stage of an insect

Harborage: an area or object that provides shelter to pests

Heat treatment: eliminating pests by raising the temperature, typically to 120–150 degrees Fahrenheit, for an extended time.

Honeydew: a sweet substance excreted by aphids and a major food source for many insects, including ants and wasps

Insect light trap (ILT): a trap with light and glue boards to attract and capture flying insects

Integrated pest management (IPM): a process or program involving detailed inspections, identifying the pests involved, employing multiple treatment methods (including chemical and nonchemicalcontrol methods), and following up and communicating those results

Larva: immature stage of an insect, generally referring to immature stages that differ greatly from the adult stage of the insect

Monitor: an insect or rodent trap used to track progress in pest control

Multiple-catch trap: a rodent trap that can catch multiple rodents

Nonresidual pesticide: a pesticide, when it dries or dissipates, that has little or no effect on pests.

Nosocomial disease: a disease contracted from a hospital that a person didn't enter with

Nymph: a juvenile insect, usually associated with insects that undergo simple metamorphosis

Pedicel: the place on ants where the alitrunk and gaster are attached that usually contains one or two nodes and are helpful in identifying ant species.

Pedipalps: the pair of appendages on the front of a spider's cephalothorax used as a sensory organ and to help capture and hold prey

Pheromone trap: a trap in which the attractant is a chemical secreted by the insects in their social or sexual nature

Predacious: predatory

Pronotum: plate-like structure covering the thorax of some insects

Pest vulnerable area (PVA): an area that is at high risk for pest activity

Pyrethroids: a class of commonly used insecticides popular because they are less toxic to people, pets, and the environment than earlier pest control chemicals

Pupa: the inactive immature stage of an insect, in between larva and adult stages

QualityPro certified: a mark of excellence given by the NPMA to companies that meet high standards in hiring practices, training, insurance, contracts, marketing, and communication

Residual pesticide: pesticide that, after the product is applied and dries, still has an effect on insects that come in contact with it; its effects may last a few weeks or up to ninety days or more—as long as the product remains in place during that time, it will kill insects that crawl over it

Sebum: oily secretion left by rodents

Simple (incomplete) metamorphosis: process some insects go through to become an adult that has three life stages: egg, nymph, and adult

Stored-product pest: a variety of insects including beetles and moths that together damage more than 10 percent of the world's grain production

Synanthropic: a wild organism that lives with or in close proximity to humans and benefits from their association and the environments humans create

Termiticide: insecticide used to kill termites

Thigmotaxis: an organism's response to the stimulus of touch or contact; positively thigmotactic organisms seek contact with objects, while negatively thigmotactic organisms avoid contact

Trophallaxis: exchange of regurgitated liquids between social insects

Ultra low volume treatment (ULV): a machine created cold fog of nonresidual pesticide droplets that float in the air to kill flying insects

REFERENCES:

Bennet, Gary W., John M. Owens, and Robert M. Corrigan. *Truman's Scientific Guide to Pest Control Operations*. Advanstar Communications, 1988.

Bennett, Gary W., PhD. *Truman's Scientific Guide to Pest Management Operations*. (Questex Media Group LLC & Purd, 2010).

Corrigan, Robert M. *Rodent Control: A Practical Guide for Pest Management Professionals*. Cleveland: GIE Media, 2001.

Eiseman, Charley and Noah Charney. *Tracks and Sign of Insects and Other Invertebrates: A Guide to North American Species*. Mechanicsburg, PA: Stackpole Books, 2010.

Fishel, Frederick M. "Pest Management and Pesticides: A Historical Perspective," University of Florida IFAS Extension, https://edis.ifas.ufl.edu/pdffiles/PI/PI21900.pdf.

Hedges, Stoy A. *Anthology: The Best of Stoy Hedges*. Cleveland: GIE Media Inc., 2001.

Hedges, Stoy A. and Mark S. Lacey. *Field Guide for the Management of Structure-Infesting Beetles* 1. Cleveland: GIE Media Inc., 1996.

Hedges, Stoy A. *Field Guide for the Management of Structure Infesting Flies*. Cleveland: GIE Media Inc., 1998.

Hedges, Stoy A. *PCT Field Guide for the Management of Structure-Infesting Ants*. Cleveland: GIE Media Inc., 2010.

Mallis, Arnold. *Mallis Handbook of Pest Control*. Edited by Stoy A. Hedges. Valley View, OH: GIE Media, Inc., 2011.

Moore, D.J. and D.M. Miller. "Laboratory Evaluations of Insecticide Product Efficacy for Control of Cimex Lectularius." *Journal of Economic Entomology 99* (December 2006): 2080–2086.

National Pest Management Association. *Urban IPM Handbook: An Integrated Approach to Management of Pests in and Around Structures.* Dunn Loring, VA: National Pest Management Association, 2006.

"Pest Management Industry Fact Sheet." NPMA. http://www. npmapestworld.org/default/assets/File/newsroom/IndustryFactSheet2013Curlnumbers.pdf.

Pinto, L.J., R. Cooper, and S.K. Kraft. *Bed Bug Handbook, The Complete Guide to Bed Bugs and Their Control.* Mechanicsville, MD: Pinto & Associates, 2007.

Pfister, Koehler, and Pereira. *Journal of Economic Entomology* (2008).

Smith, Eric H. and Richard C. Whitman. *NPCA (NPMA) Field Guide to Structural Pests.* Dunn Loring, VA: National Pest Management Association, 2007.

Thorn, Kevin. "Going to the Dogs: How We Got into the Bed Bug Dog Business . . . and How You Can." *Pest Management Professional.* September 2012.

Tripplehorn, Charles A. and Norman F. Johnson. *Borror and Delong's Introduction to the Study of Insects.* Independence, KY: Brooks Cole, 2004.